Anatomy as Spectacle

Representations:
Health, Disability, Culture

Series Editor
Stuart Murray, University of Leeds

This series provides a ground-breaking and innovative selection of titles that showcase the newest interdisciplinary research on the cultural representations of health and disability in the contemporary social world. Bringing together both subjects and working methods from literary studies, film and cultural studies, medicine and sociology, 'Representations' is scholarly and accessible, addressed to researchers across a number of academic disciplines, and practitioners and members of the public with interests in issues of public health.

The key term in the series will be representations. Public interest in questions of health and disability has never been stronger, and as a consequence cultural forms across a range of media currently produce a never-ending stream of narratives and images that both reflect this interest and generate its forms. The crucial value of the series is that it brings the skilled study of cultural narratives and images to bear on such contemporary medical concerns. It offers and responds to new research paradigms that advance understanding at a scholarly level of the interaction between medicine, culture and society; it also has a strong commitment to public concerns surrounding such issues, and maintains a tone and point of address that seek to engage a general audience.

Other books in the series

Representing Autism: Culture, Narrative, Fascination
Stuart Murray

Idiocy: A Cultural History
Patrick McDonagh

Representing Epilepsy: Myth and Matter
Jeanette Stirling

Anatomy as Spectacle

*Public Exhibitions of the Body
from 1700 to the Present*

Elizabeth Stephens

LIVERPOOL UNIVERSITY PRESS

First published 2011 by
Liverpool University Press
4 Cambridge Street
Liverpool L69 7ZU

British Library Cataloguing-in-Publication data
A British Library CIP record is available

ISBN 978-1-84631-644-9

Typeset in Iowan Old Style by R. J. Footring Ltd, Derby
Printed and bound in the United States of America

Contents

List of Figures   vi

Acknowledgements   vii

Introduction   1

1   The Docile Subject of Anatomy: Gynomorphic Waxworks in Eighteenth- and Nineteenth-Century Public Exhibitions   26

2   Lost Manhood: Turn-of-the-Century "Museums of Anatomy" and the Spermatorrhoea Epidemic   53

3   From the Freak to the Disabled Person: Anatomical Difference as Public Spectacle and Private Condition   87

4   Inventing the Bodily Interior: Écorché Figures in Early Modern Anatomy and von Hagens' *Body Worlds*   125

Conclusion   143

Bibliography   150

Index   163

Figures

1 Louis Auzoux exhibiting a medical model   7
2 Comparative exhibit of a "Giant" and a "Dwarf" Skeleton,
Mütter Museum   9
3 Anatomical Venus, Pierre Spitzner   27
4 "De foetu formatu," Adriaan van de Spiegel   28
5 "Skeletons and Parts of the Human Body Arranged on
a Plinth," Frederik Ruysch   29
6 "Anatomical Head," Gaetano Giulo Zumbo   30
7 Anatomical Venus, Clemente Susini   36
8 Cover page of Signor Sarti's *The Celebrated Florentine
Anatomical Venus*   38
9 Cover page of Jordan and Davieson's *Practical Observations
on Nervous Debility and Physical Exhaustion*   58
10 Social Hygiene Division of the War Department lantern slide   60
11 "Ignore Fake Advertisements," American Social Hygiene
Association poster   61
12 Julia Pastrana, exhibition handbill   89
13 Portrait of Fanny Mills, Charles Eisenmann   103
14 Photograph of Augustine, Paul Régnard   108
15 Coney Island Side Show banner-line   113
16 Sealo the Seal Boy, promotional poster   115
17 Jennifer Miller, promotional photograph   117
18 "Muscle Man," Valverde   131

Acknowledgements

The archival research for this book was made possible by the generous support of a British Academy Visiting Fellowship (hosted by the Centre for the Interdisciplinary Study of Sexuality and Gender in Europe at the University of Exeter), an Australian Academy of Humanities Travelling Fellowship, and the Elizabeth Crahan Fellowship at the Huntington Library. My host at the Centre for the Interdisciplinary Study of Sexuality and Gender in Europe, Lisa Downing, provided invaluable guidance and advice, as well as wonderful company. Michael Rhode at the National Museum of Health and Medicine in Washington DC patiently shared his exhaustive knowledge of the museum's archives and his enthusiasm for WWI poster campaigns. The staff of both the Huntington and Wellcome Libraries cheerfully located any number of odd archival documents during my stays there. The Ashmolean Museum kindly allowed me to make digital copies of their collection of eighteenth- and nineteenth-century handbills.

My colleagues at the Centre for the History of European Discourses at the University of Queensland have read or listened to many drafts of the papers that eventually became this book. In particular, my colleagues in the history of sexuality research programme—Chiara Beccalossi, Heather Wolffram, Alison Moore and Marina Bollinger—have been wonderfully generous with their time and feedback. The director of the Centre, Peter Cryle, has been a boundless source of inspiration and an indefatigable interlocutor about histories of bodies and sexualities in the eighteenth and nineteenth centuries. Greg Hainge, in the School of Languages and Comparative Cultural Studies at the University of Queensland, has read and commented on many early drafts of this material, and provided a constant sounding board for its development. Susan Stryker, with whom I co-organised *Bodies of Knowledge: Sexuality in the Archives* in 2007, has never failed to share her encyclopaedic knowledge of material and moments often overlooked in official histories. The Somatechnics research network based at Macquarie University—especially its director, Nikki Sullivan, also Samantha Murray and Jessica Cadwaller—have been invaluable as closely affiliated colleagues and in shaping the theoretical framework of this study. I am most grateful to Chad Parkhill, who undertook the long task of securing the permissions to publish the images in this book, and to Geoffrey White and Arun Sol for their close and careful proofreading.

Warm thanks are due, too, to the performers who were so willing to make time to discuss their work and whose wonderful performances made this research so rewarding, including Eak the Geek, Mat Fraser, and Mike Finch at Circus Oz, and most especially Jennifer Miller, who allowed me to run away and join her circus for a day. Finally, my deepest gratitude goes to Michael Bolger, who accompanied me to most of the museums and performances discussed in the following pages, and without whose infectious enthusiasm and patient photographic documentation this book would not exist.

Introduction

Visitors to any of the exhibitions advertised as "the anatomical display of real human bodies" that have toured on apparently constant rotation across the UK, the USA, Europe, South-East Asia and Australia in recent years enter a space in which the pages of a medical textbook appear to have been materialised in and as three-dimensional exhibits. With names like *The Amazing Human Body, Our Body: à corps ouvert* [The Open Body], *Bodies Revealed, The Universe Within,* and *Mysteries of the Human Body,* each of these exhibitions features a series of écorché figures (that is, bodies whose skin has been removed to reveal the internal anatomy), preserved through a process of "plastination" (in which the body's organic fluids are replaced with a clear synthetic polymer) patented by the German anatomist Gunther von Hagens, founder of *Body Worlds.*[1] Such exhibitions have proved enormously popular: *Body Worlds,* the first and still the best known, claims to have received almost 30 million visitors since its first show, in Tokyo in 1995. Despite this, the commercial nature of these exhibitions—their public display of human remains for profit—has also, and unsurprisingly, been the source of considerable controversy. In 2009, the *Corps ouvert* exhibition in Paris was closed within weeks due to negative public reaction, while at the 2002 London *Body Worlds* exhibition one spectator draped the plastinate of a pregnant woman, which also showed its foetus, with a blanket, splashing the floor around the exhibit with red paint: "A womb and a baby is such a private place. It's between the mother and the child," the protestor explained to The *Guardian* newspaper. "[A]s a parent and a human, I feel it's a sacred place that doesn't deserve to be looked at. I was looking at a kind of freak show in there.[…] It was horrendous" (Browne 9). A negative review of *The Amazing Human*

Body in the Melbourne *Age* drew the same comparison, condemning the exhibition as "little more than a freak show" (Lucas 8)—one that appealed not to a genuine scientific interest in human anatomy but to a baser and more prurient taste for the macabre.

Billboard advertisements for these exhibitions certainly play up their sensationalist elements—promising the spectacle of "real human bodies!"—and von Hagens himself has openly acknowledged using provocation as a marketing strategy: "I don't mind if you're sensationalist in your article," he told The *Guardian* in 2002. "More people will come if you are" (Jeffries 3). Von Hagens' sometimes whimsical modelling of his plastinates—turning one into a chest of drawers, fanning the flesh of another into a series of lizard frills—has hardly served to discourage the comparison of *Body Worlds* to the freak show.[2] While it is certainly possible that the scandal these exhibitions have generated is part of their popular appeal, it may well be also that these exhibitions are simultaneously so contentious but compelling because they raise a series of questions about the cultural significance and ethical treatment of bodies, to which they do not offer easy, or even especially coherent, answers. The extent to which public displays of human anatomy both fascinate as spectacle and provide a figure through which questions about the meaning of bodies can be posed is reflected in the widespread circulation of anatomical images in the public sphere. Representations of anatomised bodies can be found on television shows, in tabloid magazines, as part of public education campaigns and in commercial advertising, which reflect at once the popular interest in such images and the concerns about the propriety of their public display that characterise the reception of *Body Worlds* and other exhibitions of this sort. Popular genres of television show, like the police procedural or hospital drama, feature grisly scenes of wounded or dead, or those opened by the scalpel of the surgeon or forensic pathologist, but frame these within professional contexts—the crime scene, the laboratory, the operating theatre—thereby playing out the tension between a desire to see the anatomised body, on the one hand, and the sense that its display should be confined to closed, professional spaces, on the other.

If such images circulate so widely and are so often represented in popular contexts, it is because such a high degree of cultural significance is attributed to that representation. The bodily interior is widely understood to make visible the "truth" about that body: just as government campaigns against smoking, for instance, show smoke entering into the lungs as proof of a damage not always evident on the surface of that body, so do television shows use images of the anatomised body in order to reveal, for instance, its cause of death or to identify the disease from which it is suffering. A high evidentiary value is thus attributed to images of anatomy: it is in seeing the interior of the body that we see its truth. This understanding of the importance of the anatomical interior is very evident in *Body Worlds,* with its exhortations to "look inside" the body (exhibition handbill). In contrast to its billboard advertising, promotional literature for *Body Worlds* focuses exclusively on its educational value. The exhibition is designed "in such a way that visitors experience it much as they would a three-dimensional textbook: anatomy as the foundation of the body is laid out in an educational and elucidating fashion" (www.bodyworlds.com). Its stated objective is "to educate the public about the inner workings of the human body and show the effects of poor health, good health and lifestyle choices" (exhibition pamphlet). Similarly, the Australian promoter of *The Amazing Human Body* exhibition, Dr Wayne Castle, explained that the exhibition was intended to look "like a medical textbook" (Doble 15), and its aim, according to its handbill, was "to provide audiences with a unique and educational perspective on the inner working of the human body by viewing real human specimens," thereby enabling visitors "to make more informed decisions about its care and keeping" (exhibition pamphlet). Promotional material for the recent *Corps ouvert* exhibition in Paris also describes it as "a journey to the heart of the body, a truly educational lesson which will form a bridge between medicine and the general public" (exhibition pamphlet).

Insofar as they have been successful in promoting themselves this way, these exhibitions have often been warmly endorsed by the popular press. In its review of the 2007 *The Amazing Human Body* exhibition, for instance, *The Sydney Morning Herald* assured its readership that "as well as the general public, the exhibition will attract those in the industry keen for an incredible learning experience, such as medical students, medical researchers, sporting groups and even artists" (Doble 15). With their displays of the anatomical damage caused by our own "lifestyle choices" and "harmful habits," these exhibitions place a very explicit interpretative framework around the bodies they display, instructing the audience not only *what* to see but *how* to see it. That is, beyond the actual anatomical information these exhibitions provide about the body, they teach us to see and experience the body in a very particular way, as something that requires constant self-monitoring and for whose health we are held personally responsible. It is pre-cisely on this aspect of these exhibitions—their cultural function in producing certain ways of seeing the body and understanding our own relationship to it—that this book focuses.

In order to understand the role of public exhibitions in popularis-ing this understanding of the body and our relationship to it, it must be remembered that anatomy itself has not always enjoyed its current status as a reputable and scientific discipline. Thus, while contemporary exhibitions like *Body Worlds* call on the legitimacy and authority of anatomy to justify their exhibitory practices—framing their exhibits with explanatory labels, providing lectures on human anatomy, photographing their promoters in laboratory coats to emphasise their medical credentials—in the longer history of which these exhibitions are a part we see that in its early days anatomy itself had much to gain from the public circulation of such images. The history of popular anatomical exhibitions thus draws attention to the chequered and uneven process by which medicine acquired its current status as "[t]he most prestigious social institution in Western culture" (Gilman, *Creating Beauty to Cure the Soul* 25).

Until the mid-1800s, it should be remembered, anatomical study was the object of widespread opprobrium. Prior to 1745, when surgeons were still part of the same guild as barbers and the more respected, university-trained physicians "generally never performed manual tasks" (Lane 12), practical anatomy (that is, anatomy based on actual dissection), played a relatively minor role in Western medicine.[3] Many medical professionals in the early 1800s, writes Helen MacDonald, understood dissection as "a slow, tedious, disagreeable form of work carried out by men who had to learn to overcome the natural aversion to touching the dead" (28). It is only in the late eighteenth century that practical anatomy first began to displace the then-dominant tradition of speculative Galenic anatomy, meeting with public and professional resistance. One reason for this was that dissection was believed to defile the physical integrity of the body and was, accordingly, restricted to the bodies of the destitute and those condemned to capital punishment.[4] Anatomy was thus an extension of the penal system and itself a practice of borderline legality. Its reputation reached a nadir around the time of the infamous Burke and Hare case in 1828, when Burke was convicted of murdering fifteen indigents and selling his victims' bodies to the *Edinburgh Medical College*. For the early nineteenth-century general public, MacDonald writes, anatomists often "seemed to be monsters.... Their dealings with the dead—purchasing, grave-robbing and the like—caused public outrage and rioting against anatomy schools" (30). So disreputable was anatomy that an 1828 editorial in the medical journal *The Lancet* fulminated: "It is disgusting to talk of anatomy as a science, while it is cultivated by means of practices which would disgrace a nation of cannibals" (quoted in Richardson 131).[5] Only after the passage of the Anatomy Act of 1832, which made cadavers for medical dissection more readily available (by allowing public charitable institutions to sell unclaimed bodies, or those whose families were unable to pay for their burial, to anatomy schools), did anatomy begin to redefine its cultural position.

It is precisely over this period—from the mid-eighteenth century to the mid-nineteenth—that spectacles featuring displays of human anatomy first flourished as a form of popular entertainment. The aim of this book, then, is to examine the role such exhibitions played in establishing anatomy as an increasingly respectable and important source of knowledge about bodies, and as sites in which audiences were explicitly trained to see and understand bodies in particular ways. Although public spectacles featuring displays of human bodies have a much longer history than that with which this book is concerned, the beginning of the eighteenth century marked a definitive moment in this history in two key ways: firstly, it was at this time that a new framework, that of modern medicine, was placed around these exhibitions in a way that profoundly transformed their significance; secondly, the development of new visual technologies and forms of mass media at this time allowed for more detailed images of these bodies and for their wider circulation. Thus, although the culture of public spectacles emergent during the mid-eighteenth century might seem, on first consideration, entirely distinct from the rise of modern medicine over the same period, these contemporaneous developments were, in many ways, interrelated and mutually beneficial phenomena.

Although the history of such exhibitions is now largely forgotten, it has nonetheless played an important role in shaping modern attitudes towards the body, a history the following chapters of this book seek to recover. Public responses to *Body Worlds* and other contemporary exhibitions often treat these as unprecedented forms of popular entertainment. However, from the late eighteenth century until the turn of the twentieth, exhibitions featuring anatomical displays of human bodies—in the form of skeletal remains, teratology specimens, waxwork models and living exhibits—and marketed variously as teaching facilities for medical professionals, diversionary entertainments for the middle classes, and educational opportunities for the (upper) working classes, were consistently popular. These sites included long-running commercial enterprises, such as Rackstrow's Museum of Anatomy and Curiosities (1746–1798), Dr Kahn's Museum of Anatomy (1851–1878) and Dr Spitzner's Grand Musée Anatomique et Ethnologique (1863–c.1939). Others were professional museums attached to medical colleges but also open to the general public, including the Musée Fragonard (established in 1766 by L'Ecole Vétérinaire d'Alfort), which features human and animal remains preserved by the surgeon and anatomist Honoré Fragonard; the Natural History Museum at the University of Florence, also known as La Specola (inaugurated in 1775 and the first public science museum in Europe); the Museum of the Vienna School of Surgeons, known as Josephinum (founded in 1782); the Hunterian Museum (originally the private collection of the English surgeon and anatomist, John Hunter, opened as a public institution in 1813); the Mütter Museum of the College of Physicians in Philadelphia (established in 1849); and The National Museum of Health and Medicine in Washington DC, founded as the Army Medical Museum during the American Civil War, in 1862. While the commercial collections were largely dispersed in the mid- to late nineteenth century—some (like Kahn's) vanishing into dime museums before being lost entirely, and others (like Spitzner's) eventually purchased by public museums—the vast majority of museums attached to professional organisations remain open today.

In addition to these dedicated sites, many medical institutions occasionally functioned as or in commercial spaces, exhibiting anatomical models, human remains or living specimens, often to paying audiences. In the eighteenth century, the Bethlem Royal Hospital was opened for public tours (in which the inmates were the central attraction), while the management of the Paris Morgue dressed and posed unclaimed bodies in viewing rooms, ostensibly for the purpose of public identification, to crowds that often numbered in the tens of thousands (Schwartz 45–87). The following century, the French neurologist Jean-Martin Charcot held popular public lectures that featured his patients as living models, while the anatomical modellers Joseph Towne and Louis Auzoux exhibited their waxworks at the Great Exhibition at the Crystal Palace in London in 1851 [Fig. 1]. The statistician (and eugenicist) Francis Galton ran an Anthropometric Laboratory at the International Health Exhibition of 1884, and Alphonse Bertillon, the inventor of anthropometry, exhibited at the 1893 Chicago World's Fair. In the 1900s, Martin Courney's popular Incubator Babies exhibition at Coney Island financed pioneering developments in incubation as a new technology for the treatment of premature infants. At the 1930 Hygiene Exhibition in Germany, one of the most popular displays was the anatomical exhibit "transparent man," a "life-sized transparent model whose inner organs could be illuminated" (Hau 139). This exhibit was later purchased by the Mayo Clinic Museum in Minnesota, where it remained on public display until 1974.

Many of these medical exhibitions and museums borrowed their methods of display from those developed originally in commercial

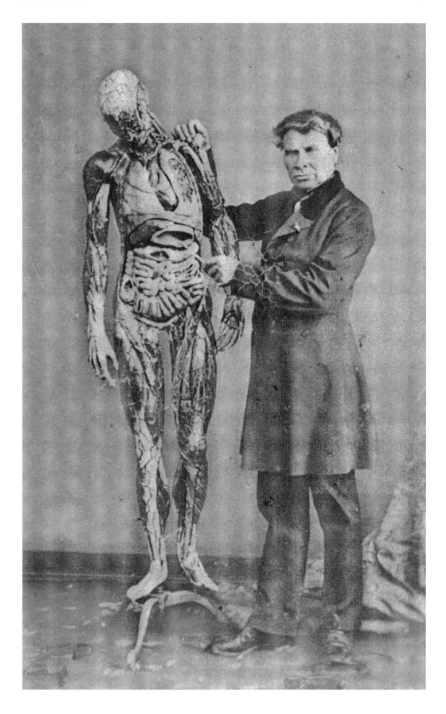

Fig. 1: The anatomical modeller Louis Auzoux exhibiting one of his figures, 1859. Courtesy of Stanley B. Burns, MD, and the Burns Archive.

contexts, such as those in popular nineteenth-century British commercial halls such as the Cosmorama Rooms, the Egyptian Hall and the Regent Gallery, where visitors could see exhibitions of new medical instruments alongside spectacles of unusual anatomies, such as conjoined twins, bearded ladies, dwarves and giants.[6] One of the central attractions in the Hunterian Museum (still on display today) is a comparative exhibit of the skeletons of Charles Byrne and Caroline Crachami, both of whom performed in late eighteenth-century exhibition halls under the stage names the "Irish Giant" and the "Sicilian Dwarf" respectively. The Mütter Museum of the College of Physicians in Philadelphia has a matching exhibition of a 7ft 6in giant displayed alongside a 3ft 6in dwarf [Fig. 2]. The manner of their display—placed side by side in the same glass cabinet, in a way that exaggerates the height of one body while emphasising the smallness of the other—exactly reproduces that of the commercial exhibition halls in which they were displayed during life.[7]

A further point of continuity between professional and popular exhibitory spaces at this time is reflected in the periodic movement of collections from private hands to the public domain and back again: the personal collection of John Hunter was originally available for viewing in Hunter's own residence, and bought only after his death by the Royal College of Surgeons in London, in 1799. It was opened as a public museum in 1813. The Musée d'anatomie, part of the University of Paris's Anatomy Department, was similarly constituted from three private collections—those of Delmas, Rouvière and Ofila—and in 1985 the institution also acquired the Spitzner collection. (This museum closed in 2005.) More recently, Robert McCoy's privately operated Museum of Questionable Medical Devices, an American archive for nineteenth-century medical instruments, was recently purchased by The Science Museum of Minnesota. As developments in imaging technologies rendered nineteenth-century collections of medical models obsolete, they once again moved into private ownership: the Liverpool School of Anatomy's collection, for instance, was sold to Louis Tussaud's museum in Blackpool.[8]

Facilitated by a reduction in working hours and the establishment of the weekend, and, with it, modern concepts of recreation and leisure, the nineteenth century saw the emergence of a wide range of new public spaces such as the public museum, the lending library, public gardens and parks, the department store, amusement parks and seaside resorts, all of which were designed to cater to a new, heterogeneous population of leisure-seekers. While the focus of these spaces was on recreation, as Andrea Stulman Dennett notes in her study of the American dime museum, there was also a strong conviction

Fig. 2: Comparative exhibit of a "giant," "normal" and "dwarf" skeleton, Mütter Museum. Courtesy of the College of Physicians of Philadelphia.

"that leisure time should not be spent in idleness and frivolity but in edifying and constructive activities" (6). The result was an exhibition-ary culture in which, as Richard Altick notes: "Scientific interest was attributed even to exhibits such as palpably contrived mermaids" (78). By the mid-nineteenth century, interest in this exhibitory culture had grown to unprecedented levels: 6 million people attended the first World's Fair in London, the Great Exhibition of 1851 (one-third of the nation's population); 32 million attended the Paris Exposition of 1889; and 27 million went to Chicago's Columbian Exposition in 1893 (half the population of the USA).[9]

In a context in which medical and exhibitory cultures were closely interrelated, and in which the current status of anatomical study was not yet established, the medical establishment initially welcomed the opening of public anatomical museums by commercial organisations. Kahn's Museum of Anatomy, which opened in 1851 (the same year as the first World's Fair), was warmly supported by the medical journal *The Lancet*, which reported: "We lately paid this exhibition a visit, and were much gratified with the collection of anatomical and surgical curiosities," commending the exhibition as "[a]ltogether a splendid scientific collection," from which "a great deal of general information is to be obtained" (unattributed, 26 April 1851: 474).[10] A follow-up article two years later remarked that, since its original review, "Dr Kahn's Anatomical Museum [...] has earned 'golden opinions from all sorts of people' [...]. The leading medical men of Scotland and Ireland have given testimonials as to its value" (unattributed, 13 April 1853: 156). In words that closely anticipate those *The Sydney Morning Herald* would use to commend *The Amazing Human Body* exhibition 150 years later, the writer continues: "The collection is well worth a visit by the profession and the general public" (156).

Anticipating the rhetoric of twenty-first century anatomical exhibitions, museums like Kahn's promised their audiences a valu-able education not only in the anatomy of the body but also in the cultivation of its health. *The Handbook of Dr Kahn's Museum of Anatomy*, like the brochures for *Body Worlds*, states that its aim was "to enable every one to become acquainted with the Laws of Health and the detrimental effects of neglecting them" (75). This emphasis on the importance of cultivating one's own health is made even more force-fully in the auction catalogue for Dr Spitzner's *Grand Musée Anatomique et Ethnologique*, which advises its readers:

Health is the happiness of life
Health is joy.
Health is strength.

Health must be everything for you.
More than money or property.
For this keep your body clean.
And keep it under constant surveillance. (*Grand Musée* n. pag.)

Exhibitions such as Spitzner's reflected a close association between anatomical knowledge, public display and health, and thus played an important role in popularising a new way of seeing and understanding the body that was then emerging. Public anatomical museums like Kahn's and Spitzner's emerged in a context of a broader exhibitory culture that, as Tony Bennett has argued, played a pivotal role in the production of new bourgeois forms of subjectivity. Bennett describes the museum as a regulatory space in which "the visitor's body might be taken hold of and moulded in accordance with the requirements of new norms of public conduct" (24), thereby "endowing individuals with new capacities for self-monitoring and self-regulation" (20) and "inducting him or her into new forms of programming that self aimed at producing new types of conduct and self-shaping" (46). Self-discipline and constant self-scrutiny were crucial to the development of the responsible and respectable bourgeois self, and nineteenth-century museums provided an important site of reinforcement for the need to undertake such practices. They

> sought to allow the people, and *en masse* rather than individually, to know rather than to be known, to become the subjects rather than the objects of knowledge. Yet, ideally, they sought also to allow the people to know and thence to regulate themselves; to become, in seeing themselves from the side of power, both the subjects and the objects of knowledge, knowing power and what power knows, and knowing themselves as (ideally) known by power, interiorising its gaze as a principle of self-surveillance and, hence, self-regulation. (Bennett 63)

The discourses of health that were simultaneously emerging in these spaces are entirely consistent with this wider bourgeois investment in notions of self-cultivation and individual responsibility. An editorial in the first volume of *The Journal of Health* in 1830 advised, typically, that:

> an attention to the preservation of health is an important duty. I do not recollect that it has often been recommended as a duty. But since our health is greatly in our own power; since we all enter into the world to engage in many active and necessary employments — and since the want of health will render us incapable of them, I cannot help thinking that the care of our health may be numbered among the duties of indispensable obligation. (unattributed 262)

This association between personal self-care and public health is argued even more directly in a popular guide for mothers written by the physician Mary Wood-Allen, who insisted: "Public health is made up of the health of individuals, and the health of grown people is very much what the children, in their care of themselves, have made it. It is the little children of today who are deciding what will be the health and vigour of the nation in years to come" (85). Here we see the way that responsibility for maximising and protecting one's health was placed firmly on the shoulders of the individual, rather than, for instance, attributed to issues of public hygiene or sanitation, or economic or environmental factors.

It is for this reason that the idea of health took on such import-ance from the middle of the eighteenth century, reinforced by a wide network of professional and popular exhibitory spaces that included public events like the Sanitation Fairs held during the American Civil War and the International Health Exhibition held in London in 1884, publications like *The Journal of Health* and *The Herald of Health and Journal of Physical Culture*, commercial products for home use like the Oxygenator or the Radium Vitaliser Water Dispenser, and sanatoria and residential retreats such as The Health Institute, The Hygeian Home and The Hygienic Institute.[11] Bruce Haley begins his study of the discursive construction of health in the nineteenth century by noting that: "[n]o topic occupied the Victorian mind more than health. Victorians worshipped the goddess Hygeia, sought out her laws, and disciplined themselves to obey them" (3). As Haley's comments and Spitzner's directives make clear, the idea of health that emerged in the Victorian period functioned very directly and explicitly as a disciplinary technology—one whose purpose, to cite Foucault's famous definition, was to forge a "docile body" that may be "subjected, used, transformed and improved" (*Discipline and Punish* 198). Popular anatomical museums can be seen to have played a pivotal role in establishing the importance of health and in encouraging popular uptake of the practices of self-care and self-cultivation that went along with it. In this way, the distinctive intermixing of the medical and the spectacular, of technologies of the body and technologies of the image, found in these museums made them important sites for training an emergent bourgeois public in how one should see the body and understanding how it should be treated, thereby interpellating a new spectator who is responsible, productive, temperate, self-disciplined, and focused on self-improvement.

As Martha H. Verbrugge recognises in her study *Able-Bodied Woman-hood: Personal Health and Social Change in Nineteenth-Century Boston*, self-government played an important role in nineteenth-century American culture because urban life itself often seemed so unmanageable:

> Sickness, personal insecurity, social change abounded; transformations in the home, the workplace, and the city strained old codes of behaviour and the familiar routines of daily life; institutions and customs that once guided conduct eroded under the force of new social conditions; a sense of order, not human perfection, was the immediate concern.[...] Self-control appeared to be the most reliable, perhaps only, mechanism for restoring order. That was the central message of health reform.[...] If disorder was the problem, personal responsibility seemed the necessary and sufficient answer. (47–48)

That the idea of health could be mobilised across this ever-widening range of contexts, acquiring increasing cultural importance as it did so, is testimony both to its capaciousness as a conceptual category and to the increasing emphasis on the importance of self-government and individual responsibility to modern subjectivities that attended it. The body was one very privileged site in which to exert the control and maintain the order which, as Verbrugge notes, was so important to nineteenth-century culture. In this context, health came to take on a very specific meaning, less designating a physical condition than naming a process, a practice of self-care and an ongoing relationship we are encouraged to develop with our own bodies. Health refers thus not simply to a particular body of knowledge, but also to a new way of seeing and experiencing our bodies as the product of the work we have put into them and the care we have taken of them.

Anatomical museums played an important role in establishing the need for self-discipline and self-cultivation by providing cautionary reminders of the grim consequences for failing to do so. As a result, as Michael Sappol recognises, even popular anatomical museums like Kahn's and Spitzner's "could plausibly claim a place in the larger project of bourgeois self-making" (294). The focus of the chapters to follow is, accordingly, not just the anatomical knowledge promoted by these popular exhibitions or their actual recommendations for the cultivation of a healthy body; rather, it is the discursive construction of anatomy and health as a set of compulsory practices of self-management and self-improvement. Recovering the history of anatomical exhibitions thus allows us to simultaneously track the emergence of the modern concept of the body as a privileged site for the cultivation of the respectable bourgeois subject, for whom work on the body comes to be seen as the privileged means by which to undertake a cultivation of the self.

By the middle of the nineteenth century, however, as the exhibitory culture of which popular anatomical museums were a part became more established, an emergent network of official or professional sites sought to secure its own position in the public sphere, by, as Tony Bennett argues, "constructing and defending that space of representation as a rational and scientific one, fully capable of bearing the didactic burden placed upon it, by differentiating it from the disorder that was imputed to competing exhibitionary institutions" (1). The by-laws of Peale's Museum (opened in 1786 and the oldest museum in the USA), provide a perfect exemplification of the way the modern museum defined itself against the surrounding commercial culture of fairgrounds and cabinets of curiosities by emphasising its order and regularity: "all articles admitted to the Museum shall be, as far as is practicable, methodically arranged, and distinctively exhibited. Each article shall be accompanied by a label, giving its scientific and vulgar name, and the locality from which it was obtained, when these are known" (*Philadelphia Museum Company* n. pag.). In a context in which order and professional control was increasingly valued, the official support enjoyed by popular museums like Kahn's underwent a radical reappraisal, reflected in the rapid decline of this museum's cultural capital in the mid-1850s. In part, this was precipitated by Kahn's collaboration, in 1857, with a "Dr" Jordan, a quack peddler of patent medicines for sexually transmitted diseases (whose career will be examined in more detail in Chapter 2). *The Handbook of Dr Kahn's Museum of Anatomy* was reprinted at this time to advertise Dr Kahn's availability for consultation "daily from 11 am till 8 pm" (front leaf) in his new medical clinic adjacent to the museum.

This move from publicising and popularising medicine to actively practising it brought about an immediate transformation in Kahn's relationship with the medical establishment. No longer simply a museum proprietor, Kahn was now considered—and forcefully denounced as—a dangerous quack, and one specialising in a particularly disreputable area of medicine.[12] Within a few short years of its earlier praise of Kahn's exhibition, the museum reappeared in the pages of *The Lancet* in a series of articles attacking that "den of obscenity [...] advertised as KAHN'S museum":

> If [...] any one is fool enough to enter its doors, he merits his fate, and richly deserves the contemptuous suspicion of his being either physically or morally diseased, which every passer-by is certain to entertain. So disgusting and immoral, so determinedly arranged for the purposes of depraving the minds of the ignorant and unwary, are the contents of this place, that their public exhibition should be suppressed by those who pretend to guard public morals and to respect public decency. (unattributed, 15 August 1857: 175)[13]

From this point, Kahn's museum became the focal point of a concerted campaign against the unlicensed practice of medicine, and particularly the treatment of sexual diseases.

The reason Kahn became such a target at this time is no doubt because he made himself so conspicuous, publicising his activities widely and occupying a prominent position on Oxford Street. For *The Lancet*, he served as the figurehead for a much bigger and more reviled industry: "KAHN is but one of a class who resort to the most unworthy means of carrying out their abominable designs," the journal asserted in a longer article several weeks later:

> By pseudo-scientific lectures, by specious advertisements, by dirty and filthy handbills [...] they contrive to carry on their unholy trade, to corrupt the morals of the young, and to prey with a relentless and vulture-like rapacity upon the victims they succeed in bringing within their influence.[...] Many an inmate of a madhouse owes his incarceration to the influence exerted upon him by the unprincipled and heartless quacks who infest the metropolis. (Unattributed. 5 September 1857: 251)

Although Kahn's museum was not closed at this time, *The Lancet*'s campaign against it and its attack on Kahn's own professional legitimacy marked a significant shift in the cultural standing and legal status of popular anatomical exhibitions as a whole, signalling a rupture of the close and mutually supportive relationship with the medical profession they had previously enjoyed.[14]

While the crackdown on Kahn's activities was in large measure a response to his move into unlicensed medical practice, it also reflects a pronounced change in professional and public attitudes towards the material on display in his museum, and the increasing sense that such exhibits were unsuited to public display. When Woodhead's Anatomical Museum in Manchester was closed in 1874, Alan Bates writes, the proprietor protested that: "the Royal College of Surgeons possesses, and admits the public to, an exhibition similar to his own" ("Dr Kahn's Museum" 622). The presiding magistrate responded, however, that "he could understand museums of the character of the defendant's being connected with the hospitals and medical colleges, but when they came into the hands of private individuals they were likely to produce serious evils" (622). The nature of these "evils" was more fully elaborated during the case against the New York Anatomical Gallery, whose proprietors were indicted for "exhibiting divers [sic] figures of men and women naked in lewd, lascivious wicked indecent, disgusting and obscene groups attitudes and positions to the manifest corruption of morals in open violation of decency and good order" (quoted in Sappol 290).

The closure of these museums was undertaken through recourse to new legislation passed in 1857 and 1858 in the UK (the USA and France would shortly follow suit). The Obscene Publications Act of

1857—passed the same year that the reportage on Kahn's museum changed so abruptly—made it illegal for anatomical museums to publish or distribute texts on venereal diseases or sexual health. Many of the handbooks of these museums (as we will see further in Chapter 2) included treatises on "marriage," whose real subject was the identification and treatment of sexually transmitted diseases and sexual dysfunction. Kahn himself was successfully prosecuted under the Obscene Publications Act in 1873 for his *Treatise on the Philosophy of Marriage* (included as a supplement to the *Handbook of Dr Kahn's Museum*). *The British Medical Journal*, covering this case, noted that Kahn was then selling 30,000 copies of his handbook per year (unattributed 151). The Obscene Publications Act was followed by the Medical Act of 1858, which regulated who could legally practise medicine in the UK and under what conditions, a move that was especially problematic for the proprietors of popular anatomical museums, who traditionally appropriated the title "doctor" without any formal—or, in many cases any—medical qualifications, and who were increasingly moving into the more profitable trade in patent medicines. By the 1860s, then, popular anatomy museums were no longer supported by the medical profession as valuable sites for the dissemination of new knowledge about the body and health management, but rather condemned as morally corrupting and physically dangerous places, to be avoided by the respectable bourgeois public. Sites of popularisation such as Kahn's were now seen as agents of vulgarisation at best, and as sites of actual malpractice at worst. Accordingly, whereas anatomical museums had originally, and successfully, marketed themselves as part of the nineteenth-century campaign for improved health and sanitation standards, by the late 1800s these venues had been re-evaluated as a part of the problem they claimed to address.

In order to understand the fate of popular anatomical museums in the second half of the nineteenth century, and why these sites of public exhibition came to be the object of the sort of opprobrium that had formerly accrued around anatomical practice, the history of these exhibitions must be placed in the context of the wider cultural transformations taking place at the same time. In the first instance, popular opinion about anatomy itself had begun to shift. Subsequent to the Anatomy Act of 1832, and as advances in both anaesthetics and sterilisation led to a series of rapid improvements in surgical outcomes, anatomical study came to be associated less with the dissection of the (criminal or stolen) dead body and more with the care of the (respectable) living. The medical profession was also being restructured and modernised, in part due to the changes resulting from the rise of anatomy as an important part of medical practice. In this context, practical anatomy consolidated its new-found respectability in the same way, as Tony Bennett argued above, that official public museums asserted their role within the nineteenth-century exhibitory complex: by distinguishing themselves from the more lowbrow and disorderly space of the fair or popular spectacle. The closure in 1855 of the annual Bartholomew Fair, which had been held in London since the twelfth century, on the grounds that it encouraged public debauchery, perfectly exemplifies the growing hostility towards such popular spectacles in the mid-nineteenth century and their gradual replacement by official events like World's Fairs (the first of which was held in London in 1851). Thus, while the crackdown on Kahn's museum was clearly precipitated by the specifics of its case and, in particular, its extension into quack medical practice, the subsequent closure of other commercial museums such as Woodhead's Anatomical Museum in Manchester and the New York Anatomical Gallery, both unaffiliated with quack clinics, reveals the extent to which this was part of larger changes in the nineteenth-century exhibitory culture and its role within the bourgeois public sphere.

While the explicit aim of *The Lancet*'s attack on Kahn's museum was, then, clearly to agitate for the line between popular and professional medical spaces to be policed more vigilantly, such distinctions should be understood as a product of the history recounted in this book rather than anterior to it. Thus, although popular anatomical exhibitions emerged at the intersection of the medical and the spectacular in nineteenth-century culture, their history is also that of the growing separation of these fields. This can be seen in the way that, during the second half of the nineteenth century, the restrictions on popular anatomical exhibitions multiplied and such exhibitions were increasingly seen, not as a mode of useful popularisation but of quackery and misinformation. We see this view articulated in Brooks McNamara's contemporary account of the popular medical shows that travelled the USA in the nineteenth and twentieth centuries. McNamara argues that the proprietors of these shows

> depended on patter, misdirection, and confusion to establish power over the crowds that assembled at his stage. His prime objective was to keep the audience interested but uncritical; he required the spectators' attention, but not such close attention that the logic of his harangue would come under too careful scrutiny. Most mountebanks found the answer in a free show which combined their lecture with tricks, demonstrations, music, and comedy. The shows had absolutely nothing to do with the medicines, cheap soap, or smelling salts hawked by the doctor. But they were entertaining and distracting, and an excellent blind for the mountebank's real object, his sales pitch to the assembled crowd. (3–4)

For McNamara, the spectacular elements of these shows served to render their audiences passive and gullible. In this sense, McNamara's understanding of the spectacle recalls that of Guy Debord, who famously described the spectacle as "the bad dream of a modern society in chains, which expresses nothing more than its desire to sleep" (11). McNamara, too, sees spectacles as lulling their audiences into a compliant stupor. However, as Armstrong and Armstrong note, while travelling medical shows did indeed conduct a brisk trade in snake oil remedies and patent tonics, some audience members bought cheap items such as soap, not because they were credulous rubes duped by the sales patter but out of politeness and gratitude for entertaining shows staged for free in remote rural locations often starved of other entertainment. Moreover, we might counter the common argument that professional museums are sites of education while commercial exhibitions are sites of trickery, by contending that it is professional museums, with their trustworthy explanatory labels and reliable exhibits, that produce passive consumers of information, whereas public spectacles like travelling medicine shows, with their reputation for (and often quite explicitly declared) humbug, teach their viewers to engage actively with the information they provide, thereby furnishing them with new interpretative skills with which to make sense of the new visual and medical technologies that were proliferating around them at an extraordinary rate during the nineteenth century.[15]

One reason this possibility has received so little critical attention may well be because the wider issue of the role of the spectacular within professional medical practice, and the extent to which ways of seeing the body are constitutive of particular kinds of knowledge about it, remains itself remarkably under-examined within histories of medicine. As Sander Gilman argues, a widespread "anxiety about the use of the visual image in the history of medicine" means that medicine has never taken proper account of the role of its own modes of visualisation in the construction of medical knowledge (*Picturing Health and Illness* 11). Accordingly, the history of medicine continues to be predominantly understood as the history of medical practice and professional organisation, while its modes and means of visualisation are considered part of the history of its popularisation rather than part of the development of its knowledge per se. Yet medical institutions have long been an important site for innovation and development in new visual technologies in ways that have transformed medical practice. During the American Civil War, for instance, the Army Medical Museum in Washington (now the National Museum of Health and Medicine) collected photographic records of case studies (including studies of gunshot wounds and amputation surgeries), publishing these in the landmark six-volume *The Medical and Surgical History of the War of the Rebellion*. These images represent early examples of professional photography and an important source of knowledge about the medical treatment of gunshot wounds.

Despite this, and in keeping with the cultural mistrust of popular spectacles that intensified during the mid-nineteenth century, the medical profession has often reacted with hostility to the public circulation of such images. The professional reception of the surgical films of the French surgeon Eugène-Louis Doyen is a case in point. Doyen recorded dozens of operations at the turn of the twentieth century—including the separation of the conjoined twins Doodica and Radica Neik, who had previously travelled with the Barnum and Bailey Circus—which he screened with commentary at international professional conferences and, to a larger audience, at the Paris World Exposition in 1900. While Doyen intended his films primarily for a medical audience, José van Dijck notes that one of his camera operators sold a copy of the film of the Neik sisters to an "impresario," who screened it without Doyen's consent in commercial spaces such as coffee houses and at fairs. Although Doyen later restricted the screening of his films to medical schools and other educational spaces, Doyen himself became the object of professional censure as a result: "One of his colleagues, Dr. Legrain, surgical chief at the Ville-Évrard hospital, publicly accused Doyen of charalatanism and clowning," van Dijck writes, while medical schools proved uninterested in screening his films to their students (154).

This association of the spectacular with the quackish reveals a deep institutional distrust within the medical profession not only about the popularisation of medical knowledge but more especially about its commercialisation, reflected in long-standing bans by most national medical associations on their members advertising medicines or medical treatments. As early as the late eighteenth century, Samuel Solomon tried to challenge the logic underlying this decision, writing in *A Guide to Health; or, Advice to Both Sexes*:

> It cannot be presumed that the mode of making known the virtues of a medicine, through the medium of a Newspaper, is even *improper, much less disgraceful*. If the REMEDY be of SUPERIOR EFFICACY, it becomes a duty to the public, and to himself, that the inventor or proprietor do make it as generally known as possible; and it must not be argued that the MEDICINE is *debased*, by the channel through which society are made acquainted with it. (xii–xiii)

Solomon's argument was rather compromised by the fact that he himself was a particularly high-profile empiric, inventor of the popular but misleadingly advertised Cordial of Gilead.[16]

Those who followed Solomon's enterprising lead were also either quacks or popular health reformers. In the nineteenth century, the temperance advocate Sylvester Graham, inventor of the Graham cracker and wholegrain enthusiast, was the first to reach a mass audience through public lectures and "the first to publicise his cause in the new popular press" (Armstrong and Armstrong 55). Several decades later, the diet reformer John Kellogg was among "the first to make use of nationwide mass-market advertising [...] at a time when few foods of any kind were heavily advertised on a national scale" (Armstrong and Armstrong 111). In the twentieth century, the "rejuvenation doctor" John Brinkley (who claimed to restore virility by transplanting goat glands into human testicles), invented commercial radio after buying his own station as an advertising medium for his patented medical formulas, on which he hosted the hugely popular "Medical Question Box."[17] Brock notes that Brinkley happily accepted advertising money from rival medical entrepreneurs, broadcasting advertisements for a wide range of toxic products, including "Kolorbak hair dye (which caused lead poisoning), Lash Lure (blindness), Radithor ('Certified Radioactive Water'), and Koremlu, later described by investigators as 'a depilatory made from rat poison'" (180–181).[18] Despite the dubious quality of their products and their profit-driven motivation, medical promoters in the nineteenth century did much to popularise an increasingly anatomised view of the body to a general public, thereby contributing to the growing dominance of medicine as a privileged source of knowledge about the body and its care.

Accordingly, the change in status experienced by popular anatomical museums during the mid-nineteenth century needs to be understood within the context of the wider changes in the relationship between visual and medical cultures taking place at this time across a wide range of fields that were all associated, to a greater or lesser degree, with the nineteenth-century exhibitory culture. Most significantly, it is only in the mid- nineteenth century that medicine first became invested in the idea that anatomical knowledge could be represented in an objective way, free from cultural mediation, and that the scientific and the spectacular came to be seen not only as fundamentally different to one another but also as categorically opposed. (Not coincidentally, this is also the period in which photography emerges as a new visual technology, claiming to record images in a direct and unmediated way.) The publication of *Gray's Anatomy: Descriptive and Surgical Theory* in 1858 exemplifies this shift: whereas earlier styles of anatomical illustration foregrounded their use of artistic convention, seeing cultural commentary on the body as a central part of anatomy's purpose (as we will see in the following chapter), *Gray's Anatomy* signalled an epochal change in the visual codes used in medicine, representing a shift towards the diagrammatic and images decontextualised from cultural referents. In her study of nineteenth-century medical photography, Amy Werbel refers to this period as one of the "neutralisation of the image," characterised by a sort of new "non-style" that was, nonetheless, "itself a very special mode of rhetoric, the rhetoric of disinterested objectivity. Medical photography, in short, evolved a style most notable for what it did not have, namely aesthetic lighting, props, and narrative content" (103). We see the consequences of this shift in the way medicine has, since the mid-1800s, increasingly marginalised the role played by visual images in the production and circulation of medical knowledge, assuming these to serve a purely illustrative purpose, that is, to represent, rather than construct, knowledge about the body.

For Jonathan Crary, however, the new visual technologies that started to appear in the early nineteenth century represented "far more than simply a shift in the appearance of images and art works, or in systems or representational conventions. Instead, [they were] inseparable from a massive reorganisation of knowledge and social practices that modified in myriad ways the productive, cognitive, and desiring capacities of the human subject" (3). Catherine Waldby, too, has argued that the history of medical imaging technologies is not that of an increasingly detailed and accurate representation of a pre-existing body, but rather a process of actively transforming those bodies and the way that they are seen, continually remaking them in its own image (227–43).[19] For this reason, she contends, the use of visual technologies needs to be reconsidered as integral rather than external to the development of medical knowledge. Against a long history that views spectacles and images as merely the means by which information is (mis)communicated, this book seeks to draw attention to the way visual technologies and sites of public spectacle actively produce new knowledges and new ways of seeing, thereby constituting the objects of knowledge they claim merely to reveal.

The marginalisation of the visual conventions of anatomy is evident even in contemporary exhibitions like *Body Worlds* and *The Amazing Human Body*, despite the fact that these are clearly very much part of its spectacular tradition. The claims by Wayne Castle and Gunther von Hagens that their shows represent human anatomy "like a medical textbook"—as though anatomy textbooks themselves depict anatomy in an objective or unmediated way—exemplify the assumption that the medical and the spectacular are mutually exclusive possibilities, even as they problematise it. In treating their spectacular and medical elements together, this book seeks to show how such public exhibitions represent anatomy not simply as a particular field of knowledge about bodies but also, more importantly, as a lens through which to see those bodies, through which the cultivation of one's body came to be understood as a social obligation and as a path to personal fulfilment. The contexts in which medical and spectacular cultures have come together to provide a new way of seeing and knowing bodies since the emergence of practical anatomy and the public sphere in the early eighteenth century are so varied that recovering this history requires moving between very different kinds of texts, cultural spaces, and orders of knowledge. Accordingly, while the exhibitions considered in the following chapters all represent key moments in the spectacular display of human anatomy, they are not intended to comprise an exhaustive account of anatomical exhibitions. Rather, they are designed to show how such exhibitions have trained us to see work on the body as both a personal responsibility and a privileged path to self-realisation and personal fulfilment.

This book, then, seeks not simply to recover a now largely forgotten episode in the history of medicine and to show its ongoing impact on popular understandings of bodies. Rather, it is also concerned to show how these exhibitions (re)produced an emergent anatomical vision of the body, a new mode of seeing that body that constructs, rather than simply represents, new kinds of knowledge about it. As we will see in the chapters to follow, public exhibitions of human anatomy have historically become most popular precisely during those periods in which the cultural significance of that body is most unstable. Following Marina Warner's insight in *Fantastic Selves, Other Worlds*, that images of corporeal transformation tend to proliferate during periods of rapid cultural change, the following chapters will show how interest in public exhibitions of anatomy intensifies during those moments in which that anatomy's significance is being actively challenged or transformed; and they will offer new guidelines for seeing and new rules to follow. Thus, these exhibitory spaces are not sites in which dominant ideas about bodies are reproduced or simply transmitted; rather, they are sites at which new ideas about bodies are both formulated and contested. In these spaces, it is important to recognise, dominant discourses shape but do not determine the significance of the bodies on display, which derives instead from the volatile and always dynamic interrelationship between the spectator, the exhibit, and the space of exhibition.

Notes

1 Although none of these exhibitions are formally affiliated with *Body Worlds*, whose website refers to them as "plagiarists," all appear to source their exhibits from the same processing facilities. Given the limited number of such facilities, and the fact that von Hagens owns the patent on plastination, it is likely they are all associated with von Hagens to some degree.

2 Despite this, there are significant differences between the exhibitory style of *Body Worlds* and the freak show: in the first place, freak shows primarily display living models rather than human remains; secondly, they rarely advertise their exhibits as morally improving or instructive; and, finally, freak shows focus on figures of anatomical difference—"JoJo the Dog-Faced Boy," "the Hottentot Venus," "the Elephant Man" etc.—whereas contemporary anatomical exhibitions predominantly feature displays of "normal" anatomy. In light of these differences, I would suggest that the reason it is the freak show that is so often invoked by critics of these exhibitions is because the freak show is now the only reference point to the longer history of popular anatomical displays with which contemporary audiences are likely to be familiar.

3 As Jonathan Sawday notes in *The Body Emblazoned: Dissection and the Human Body in Renaissance Culture*, the "anatomical Renaissance" can be traced to the sixteenth and seventeenth centuries in Europe, the period in which the recovery of Greek medical texts and the emergence of dissection transformed medical understanding and treatment of the body (39–43). However, as Barbara Stafford (49) recognises, it was only at the beginning of the eighteenth century that "anatomy became the basic science for surgeons," and that surgeons began to distinguish themselves from the barbers with whom they were associated by guild, and to acquire the kind of prestige that had previously been reserved for physicians.

4 Detailed histories of popular attitudes and legal regulations regarding dissection and anatomy can be found in Sawday, MacDonald, and Ruth Richardson's *Death, Dissection, and the Destitute*. As Katherine Park has argued, however, the history of practical anatomy and its public reception is more complex than most histories recognise and begins much earlier. Park's own account, in *Secrets of Women: Gender, Generation, and the Origins of Human Dissection*, concentrates primarily on the twelfth and thirteenth centuries, during which time autopsies were undertaken as part of a range of religious and other cultural practices that had little to do with medicine and extended to the highest echelons of society (such as aristocratic women who had died in childbirth).

5 Burke's body was itself donated to the medical school at Edinburgh after execution, and his skeleton is still on display at the University Medical School. A pocket book was subsequently made of his skin and this is on display at the Police Museum on the Royal Mile.

6 Similarly, spectacles in commercial contexts often laid claim to medical or scientific purpose. The phrenological exhibition run by The Fowler Brothers (authors of a series of popular texts on phrenology, as well as editors of the American Phrenological Journal) was one of the most popular attractions in New York in the 1840s: "While the crowd slowly wandered through the museum, the more adventurous stood in line for a phrenological reading by one of the famous Fowler brothers," Armstrong and Armstrong write. "[T]he brothers were true entertainers …. They whipped out thick tape measures for measuring each customer's head; read heads while blindfolded; took turns examining the same head while the other brother left the stage" (71–73).

7 The means by which the bodies of Byrne and Crachami were acquired by medical authorities raises a number of important questions about the ethics of official medical museums and whether these differed historically in any significant way from those of freak shows and other commercial spaces. Byrne was so resistant to the idea of medical dissection and posthumous display that prior to his death he paid two men to sink his body in the Thames; they nonetheless sold his remains to the Royal College of Surgeons. Crachami, who was nine years old at the time of her death, had her autopsy interrupted by her father, who had not given consent and was assured that the procedure would be definitively halted (see Youngquist 5–9).

8 This museum still exists, now as part of the Ripley chain, although this earlier material is no longer on public display.

9 As Julie Brown has recently shown (*Health and Medicine on Display*), displays focusing on sanitation and health issues were an important focus of international expositions in the USA throughout the nineteenth century, including artificial anatomical models, prostheses, surgical procedures, statistics and graphs on epidemic diseases and sanitation programs, models of housing developments, and so on. The use of live models to exhibit congenital disorders and their treatment appears not to have been a feature of these expositions.

10 Kahn's claim to a medical title is somewhat uncertain. Although Karl Pearson identifies Kahn as a qualified medical doctor who had taught at the University of Vienna, he provides no source for this information. Kahn was later tried (in January 1874) on charges of falsely representing himself as a doctor, as he was not registered in the United Kingdom, but he was acquitted.

11 This preoccupation with health developed, it should be noted, in a period marked by a series of epidemic diseases that saw life expectancy lower in the nineteenth century than in the eighteenth (Hall and Porter 125–31).

12 In the nineteenth century, Angus McLaren notes: "much of the odium connected with venereal disease extended even to the medical men who treated it" (127).

13 Karl Pearson suggests that the actual exhibits in Kahn's museum collection may also have changed at this time, although the museum's handbooks provide little evidence of this. Pearson claims that the museum was probably originally "what it professed to be—a museum for the study of anatomy—but its proprietor soon found that the shillings rolled in from an inquisitive lay public, and accordingly Dr Kahn started introducing monstrosities of all types approaching near to those of the showmen at village fairs" (214).

14 Despite the temporary closure of Kahn's Anatomical Museum in 1857, it was not until 1873 that the Society for the Suppression of Vice was finally mobilised by the police to seize Kahn's anatomical models, which Karl Pearson notes were destroyed at the Malborough Street Police Court on 18 December 1873. (Although Alan Bates argues that the collection was shipped to New York City and reopened in the Bowery District. ("Dr Kahn's Museum" 623)).

15 Such skills were especially important during a period in which "the father of American medicine," Benjamin Rush, recommended treatments that included "bombing the body with mercury-laced calomel (which caused rampant diarrhoea, bleeding of the gums, and uncontrolled drooling), blistering with hot irons, tobacco-smoke enemas, and bleeding by the pint" (Brock 11). In the early 1800s, American doctors also discouraged the consumption of vegetables, believing this a contributing factor to the cholera epidemic (Armstrong and Armstrong 54), while "dirt was [considered] something positive, even healthy" (Hoy 3), a preference that would only give way after persistent advertising

campaigns by chemical companies manufacturing new products such as toothpaste and deodorant.

16 Solomon (1768–1819) received an MD from Marischal College in 1793, although he was suspected of forging his certificates from two local doctors. Although he promoted his cordial as having no other active ingredient than "Gold! pure virgin Gold!" (36; original emphasis), his formula was later found to consist primarily of brandy (*Journal of Health*, "Quackery" 349).

17 As Pope Brock notes, Brinkley also pioneered the use of pre-recorded broadcasts, and transformed then-unknown hillbilly and country performers, including the Carter Family, into stars.

18 False claims about the contents and efficacy of patent medicines did not become illegal in the USA until the establishment of the Food and Drug Act in 1907, which made the listing of actual ingredients compulsory. The establishment of the FDA was a direct result of the investigative work of writers such as Anthony Comstock, who recounted his findings in *Frauds Exposed* (1880) and Samuel Hopkins Adams, whose *The Great American Fraud* (1906), based on his groundbreaking series of articles for *Collier's* magazine, is widely recognised to have directly led to the establishment of the FDA the following year.

19 For instance, many of these technologies produce digitised scans of the body that are later translated into visual images via complex computer programs, resulting in highly processed rather than transparent images of the body.

CHAPTER 1

The Docile Subject of Anatomy: Gynomorphic Waxworks in Eighteenth- and Nineteenth-Century Public Exhibitions

By the time Paul Delvaux discovered "Dr" Spitzner's anatomical Venus [Fig. 3], exhibited in Spitzner's Grand Musée d'Anatomie et d'Hygiène during the 1932 Foire du Midi de Bruxelles [Brussels Midi Fair] in 1932, the popularity of such anatomical exhibitions, which had reached its peak in the mid-1800s, had entered a period of terminal decline.[1] The history of Spitzner's own collection exemplifies the transformations such museums underwent over the course of the late nineteenth century.[2] Having first opened on a dedicated site in the centre of Paris in 1856 as Spitzner's Grand Musée Anatomique et Ethnologique, the museum became itinerant in the late 1800s, and by the century's end was to be found in the fairgrounds of Paris, before closing definitively around the time of the Second World War.[3] The paintings Delvaux produced inspired by this collection, *La musée Spitzner* (1943) and *The Sleeping Venus* (1944), are considered among his most important works (Scott 60), and attest not only to an enduring popular interest in these exhibitions, which long outlasted their heyday (or even relevance) as scientific collections, but also to the way representations of anatomy have a cultural resonance that extends beyond the "purely" medical. Delvaux's appreciation for Spitzner's Venuses was aesthetic rather than scientific, and was reinforced rather than undermined by its anachronism as a medical model.

Spitzner's Venus, with its highly aestheticised representation of female reproductive anatomy, may appear very odd and unscientific to contemporary eyes. However, this medical model, designed to demonstrate the then-new procedure of the Caesarean section, was made at the end of a long period in which art and anatomy were closely interrelated disciplines, which, as we saw in the previous chapter,

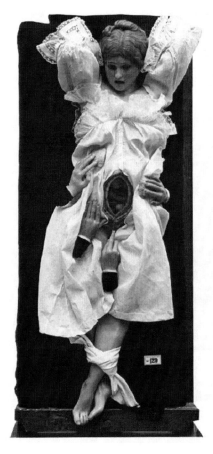

Fig. 3: Spitzner's Anatomical Venus (date unknown). In *Le Grand Musée Anatomique et Ethnologique du Dr P. Spitzner* (auction catalogue). Courtesy of the Wellcome Library.

only came to an end in the mid-nineteenth century (around the time this model was manufactured). Although this tradition is now largely forgotten, prior to the middle of the nineteenth century anatomical models and images routinely drew on visual conventions now associated with artistic representations, and saw explicit commentary on the cultural significance of the bodies depicted as central to its purpose: seventeenth-century anatomical artists like Spigelius represented dissected bodies as part of natural or architectural landscapes [Fig. 4]; the anatomist Frederik Ruysch transformed the skeletons of dissected infants into moral tableaux [Fig. 5]; while many early models of artificial anatomy have a distinctly melancholy air, or are depicted as wide-eyed and stunned by the fact of their dissection. Gaetano Giulo Zumbo's "Anatomical Head," often cited as the first example of an

Fig. 4: "De foetu formatu," Adriaan van de Spiegel (Spigelius). In *Opera quae extant omnia* (1645). Courtesy of the Wellcome Library.

anatomical waxwork made especially for medical study (de Ceglia 2), and still on display at La Specola, the Natural History Museum at the University of Florence, is a prominent example of this [Fig. 6]. Prior to Zumbo, Haviland and Parish note, such models of the human form were made by artists rather than anatomists, and "were used for teaching anatomy in art schools, not medical schools" (54). It is in this context that anatomical Venuses like Spitzner's need to be understood.

In addition to the history of medicine, anatomical Venuses also need to be contextualised within the history of public exhibitions. While part of the appeal of Venuses as objects on public display no doubt

Fig. 5: "Skeletons and Parts of the Human Body arranged on a Plinth," Frederik Ruysch. In *Thesaurus Anatomicus* (1703). Courtesy of the Wellcome Library.

derived from their novelty factor, their popularity would also have been dependent on their legibility as aesthetic objects, their ability to be understood through established conventions for public display. Exhibitions of Venuses had two main precursors which would have been widely familiar to eighteenth-century audiences and thus would have enabled audiences to make sense of their public exhibition. The first was the very long tradition of displaying waxwork statues as objects of art or religious observance in public places such as churches

Fig. 6: "Anatomical Head," Gaetano Giulo Zumbo (1700). Postcard. Courtesy of the Natural History Museum at the University of Florence.

and cabinets of curiosities.[4] The second was the tradition in the visual arts of representing female nudes as Venuses, of which the Venus de Milo, Botticelli's *Birth of Venus* and Titian's *Venus of Urbino* are some of the better-known examples.[5] The existence of these earlier traditions, and their incorporation into the modelling of anatomical Venuses, at once explains something of their popularity and the ease and speed with which they were able to move from the professional medical context into the public sphere. Thus what the modelling of anatomical Venuses reveals is both how anatomical exhibitions came to be circulated in the public sphere, thereby medicalising popular ways of seeing the body, and also the ways anatomical knowledge itself was shaped by cultural conventions for representing such bodies and popular assumptions about their significance.

My aims in this chapter are thus threefold. Firstly, to trace the history in which anatomical Venuses were originally manufactured for a professional clientele but quickly came to be exhibited to the general public, drawing attention to the close interrelationship between the spheres of professional medical practice and popular exhibition to which this history attests. Secondly, to consider what the popularity and the changing aesthetics of these models has to tell us about contemporaneous transformations in the cultural conceptualisation of femininity and maternity over the period in which they were manufactured and exhibited. Thirdly, and coextensively, to take account of the role these spectacular models of female reproductive anatomy played in the increasing public acceptance, and eventual dominance, of anatomy as a privileged source of knowledge about the body and its care. Public exhibitions of anatomical Venuses encouraged viewers to see themselves both as docile subjects of health and medicine, but also as responsible and productive citizens prepared to educate themselves about, and to accept responsibility for, the health of their bodies, along with those of their families and their communities. In examining the transformations in the aesthetics and exhibitory contexts of Venuses during the eighteenth and nineteenth centuries, we can clearly see the extent to which these (re)produce contemporaneously changing ideas about the body, demonstrating how popular assumptions about bodies are informed by the official knowledge about those bodies, and how scientific knowledge is in turn shaped by prevailing cultural values.[6] However marginal these models may now seem to the history of medicine and popular culture, as we will see, they constitute an important site for the emergence of the new bourgeois subject — one that is docile, self-cultivating, responsible for his or her own health.

Venus Anatomised:
The Birth of the Anatomical Museum

Anatomical Venuses were among the first artificial anatomical models ever produced. Artificial anatomy itself emerged as a specialised branch of medicine during the very late seventeenth century, and is usually dated to the work of Gaetano Giulo Zumbo (1656–1701).[7] Manufactured from the early 1700s onwards, and originally intended for use as teaching models, anatomical Venuses were designed to provide detailed and accurate waxwork models of the female reproductive system, including various stages of foetal development and childbirth.[8] This was a period in which legal regulations and lack of refrigeration combined to produce a shortage of available cadavers for dissection and anatomical instruction.[9] Female cadavers — especially pregnant cadavers — were particularly difficult to find, because, as Helen MacDonald writes: "It was rare for a woman to hang for murder in England at this time, so of the bodies received at the [Royal] College [of Surgeons] between 1800 and 1832, only seven were female" (19). Pregnant women condemned to public execution, whose bodies were thus available for dissection, could "plead their bellies," and be granted a stay of execution until after the birth of their child (such women were often subsequently pardoned). Puccetti notes that, in consequence, gynaecology did not develop as a medical discipline until the eighteenth century, prior to which childbirth was considered a "natural event outside the purview of medicine" (80), so that doctors in the eighteenth century "had scarce knowledge of the anatomy of female genital organs and of the physiology of delivery" (80).

Anatomical Venuses were designed to remedy this gap in medical knowledge. As such, they reflect the increasing importance of gynaecology as a specialised branch of medical research and the development of its practice at this time.[10] Despite being originally intended for a professional audience, however, Venuses began to circulate almost immediately in the popular sphere, touring as public exhibits marketed to a general audience from the early 1700s onwards. Indeed, the very first known example of an anatomical Venus almost immediately found its way into public display, attesting to the productive interrelationship between the simultaneously emergent fields of artificial anatomy and public exhibitions, and between professional knowledge and popular spectacles, at this time. Public interest in these exhibitions was substantial enough that when the first permanent museums of anatomy were established in the middle of the eighteenth century, Venuses were invariably advertised as their star attractions. Early examples of Venuses can hence be seen to have cultivated a public taste for popular anatomical exhibitions and so to have played an important role in popularising and disseminating the new ways of seeing and understanding the body these produced. In order to make sense of the exhibition of Venuses, and to examine what this history has to tell us about changes in the cultural status of anatomy and the implications of this for the way bodies were popularly seen and understood in the eighteenth and nineteenth centuries, this part of the chapter will begin by recovering an account of their manufacture and the circumstances of their rapid movement into the space of public exhibition, grouping these into three distinct generations.

One of the earliest pioneers in artificial anatomy was the French surgeon and professor of anatomy Guillaume Desnoües (c. 1650–1735), who worked briefly in collaboration with Zumbo.[11] Desnoües' first completed model is believed to have been that of an anatomical Venus which, as would become standard, also included a model of the foetus (Haviland and Parish 56). Shortly after exhibiting his figure to the French Académie des Sciences in 1711, Desnoües opened a public museum of anatomical waxworks in Paris. He then moved to London in 1719. The general public, it would appear, was more immediately responsive to his models than the professionals for whom they had been originally intended: it was only after several years of exhibiting his work in commercial rooms in the UK and France that Desnoües' work began to receive attention from the medical community, enabling him to hire his figures out to anatomists and surgeons for the teaching purposes for which they had been designed (Haviland and Parish 57).[12] In the late 1720s, Desnoües' models were used in lectures by surgeons offering private courses in anatomical instruction (reflecting the gradual rise of practical anatomy as an increasingly central part of medical training and practice).[13]

Almost no trace of Desnoües' Venuses remains: there are no surviving models, or images from catalogues, or written descriptions. In a rare recorded piece of correspondence, Desnoües explained that his work was intended to teach popular audiences about anatomy "without exciting the feeling of horror men usually have on seeing corpses" (Haviland and Parish 56). As Maritha Rene Burmeister argues in her study of nineteenth-century popular anatomical museums: "For anatomical Venuses on public display, one of their most important features was their ability to refer directly to medical science but in a manner devoid of corporeal reality" (51).[14] Desnoües' comment, brief as it is, hence helps us to place the earliest Venus in its cultural context, as a professionally valuable medical model and as an object suitable for public display. Moreover, it reveals that Desnoües needed to teach spectators how to interpret its meaning: instructing audiences not only on what they were seeing (the internal organs of the female reproductive system) but also on how they should see it.

Although there are no records of attendance figures at the exhibitions of Desnoües' Venus, the frequency and duration of its career as a touring exhibit during the first half of the eighteenth century provides an indication of its popularity, as does the fact that by 1733 a second early Venus, made by another modeller, had joined it on public display. Manufactured and exhibited by the medical demonstrator Abraham Chovet (1704–1790), this Venus, like Desnoües', was originally intended for use in professional medical institutes as a demonstration model — specifically for Chovet's lectures on anatomy at the Surgeon's Hall in London from 1734 to 1736.[15] Like Desnoües' Venus, Chovet's quickly found its way into the world of commercial exhibitions. The actual model of Chovet's Venus, again like that of Desnoües, has now been lost; however, a detailed description of Chovet's Venus is given in the pamphlet produced for its exhibition. Chovet's Venus was advertised as a mechanical marvel as well as an anatomical model, one which made visible "the Circulation of the Blood [...] through Glass Veins and Arteries, with the Actions of the *Heart and Lungs*; As also, the *Course of the Blood* from the Mother to the Child, and from the Child to the Mother" (*An Explanation of the Figure of Anatomy*, frontispiece). The catalogue for Chovet's Venus emphasised its educational value for the general public: "any Person, tho' unskilled in the Knowledge of ANATOMY, may at one View be acquainted with the *Circulation of the Blood*, and in what Manner it is performed in our living Bodies" (frontispiece). At the same time, this lesson in human anatomy was presented to the public in a style that appears to have been spectacularly gory:

> As this Figure is chiefly calculated to demonstrate the *Circulation of the Blood*, with the Actions of the *Heart* and *Lungs*, and the Nourishment of the Child while in the Womb, it was absolutely necessary that it should represent a Woman, supposed to be opened when alive; because these are all the vital Functions, which are not exercised in a Body when dead. This Figure represents a Woman gone eight Months with Child, chained down upon a Table, supposed to be open'd alive, of which the two principal Cavities are laid open. (*An Explanation of the Figure of Anatomy* 3)

Chovet, like Desnoües, took pains to remind his audience that his exhibit was an ingeniously made *model* and not a real body, claiming "it is to be hoped that nobody will make objection to the Representation, which would carry with it an Idea of the highest Barbarity and Cruelty, had it ever been put in Practice upon any Humane Body" (*An Explanation of the Figure of Anatomy* n. pag.). Given that Chovet's Venus is depicted in the midst of vivisection rather than dissection, it might justly be argued that his figure is represented as being subjected to a practice much *more* "barbaric and cruel" than that used in actual anatomical study.

It is precisely this aspect of Chovet's exhibition—as described in his catalogue, his sensationally grisly display of female reproductive anatomy—which makes it so historically informative, elucidating not only the state of anatomical knowledge and gynaecology at this time but also its cultural significance. Straining against the ropes that bind her to the table, represented as "open'd alive," Chovet's Venus is reflective of a culture in which dissection is understood as a disciplinary spectacle, a punitive act that reinforces the eighteenth-century association between anatomy and the penal system.[16] Thus while Burmeister argues that Chovet's Venus was exhibited to an audience who evidently "did not demand the removal of any reminders of suffering as a condition for viewing the models" (34)—that is, who were more accustomed than twenty-first century audiences to spectacles of violence or death—nonetheless the exhibitory style of this Venus seems to reflect a negative public opinion about dissection and a sense of dread and disapproval about its practice. However, it does not appear that the atmosphere of scandal and violence that accrued around dissection at this time hindered public interest in its exhibition. On the contrary, the rhetoric of Chovet's catalogue suggests it was precisely its mixture of the disciplinary (or educational) and sensational (or gory) that made this spectacle so fascinating for eighteenth-century audiences.

That the Venuses of both Chovet and Desnoües were popular objects of public display is further evidenced by the fact that both were later bought by Rackstrow's Museum of Anatomy and Curiosities, where Chovet's remained on public display until the museum's closure in 1798. (The Desnoües Venus was sold in 1753.)[17] Rackstrow's, the first permanent popular anatomical museum in the UK, exhibited artificial anatomical models and real human skeletons alongside a typically heterogeneous eighteenth-century collection of objects that included Egyptian mummies, the hide of a rhinoceros, an armadillo, a porcupine, two crocodiles, a pair of Laplander snowshoes and a quiver of Indian arrows (*A Descriptive Catalogue (Giving a full Explanation) of Rackstrow's Museum*). Although its curatorial style was clearly shaped by the earlier tradition of cabinets of curiosities, Rackstrow's catalogue reveals that anatomical exhibits were, from the outset, central to the museum's collection, vastly outnumbering its other exhibits: the 1784 catalogue lists ten Venuses among the museum's 117 listed objects on display, suggesting the extent to which these were recognised as key to the popular interest in such exhibitions at this time.[18] Thus the length and frequency of their public exhibition suggests that the Venuses of Desnoües and Chovet did much to foster a popular interest in anatomy during a period in which, as we have already seen, anatomy was still largely held to be a disreputable discipline.

By the close of the eighteenth century, however, public interest in anatomical exhibitions was evidently waning. After the closure of Rackstrow's in 1798, there are no records of subsequent exhibitions of Venuses until the early 1840s, when a second generation of models began to be toured across the UK, again meeting with apparent critical and commercial success.[19] The best known of these figures were the Florentine and Parisian Venuses, named after their respective places of origin.[20] While the manufacturer of the Parisian Venus is not documented, several well-known anatomical modellers were producing full-body waxworks in Paris in the first part of the nineteenth century: the catalogues of Jules Talrich and Docteur Auzoux both include Venuses in their list of figures for sale. The Florentine Venus, however, was almost certainly made by Clemente Susini (1754–1814),

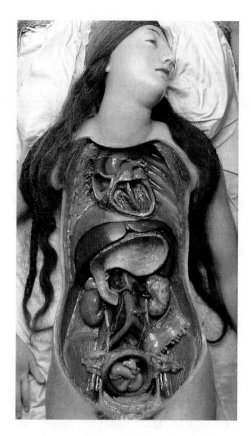

Fig. 7: Anatomical Venus by Clemente Susini (c. 1798). Postcard. Courtesy of the Natural History Museum at the University of Florence.

who worked in the anatomy laboratory of La Specola (the Natural History Museum at the University of Florence) from the end of the eighteenth century and is widely recognised as the foremost modeller of anatomical Venuses [Fig. 7].

Susini's Venuses occupy a pre-eminent position in the history of anatomical Venuses, partly because they now constitute virtually the entire archive of figures that have been preserved, most of which remain on public display (at La Specola in Florence, the Josephinum in Vienna and the Museo di Anatomica Umana in Bologna). Requiring as many as 200 cadavers to make the wax moulds used in their manufacture, Susini's figures provide extremely detailed models of the internal female reproductive anatomy: each Venus has a panel at the torso which can be detached to reveal the internal organs, removable in progressive layers until one reaches the womb and its developing foetus.

Susini's Venuses are as striking aesthetically as they are technically. Although these models were, like those considered above, originally manufactured for a clientele of medical practitioners, from the outset all Susini's Venuses incorporated ornamental details unrelated to their function as anatomical models: each is laid on a while silk sheet draped over red velvet pillows; each is posed with her back arched, mouth open, legs slightly bent, with long loose hair falling at her side and a string of pearls about the neck.

While the date of their earliest exhibition to the general public is not known, Susini's figures were quickly purchased for public museums: the Josephinum in Vienna and the Museo delle Cere Anatomiche in Cagliari both bought large early collections of Susini's work.[21] While this suggests a continuity of interest in the public exhibition of Venuses from the eighteenth to the nineteenth century, the radically different styles of their modeling also points to significant changes. The aesthetics of the Venuses of the 1840s are as far from Chovet's as one could imagine: ecstatic rather than agonised, pampered rather than suffering, represented in passive postures and domestic settings rather than bound and struggling in the public anatomy theatre.[22] In contrast to the catalogue description of Chovet's Venus as "open'd alive," the cover illustration for Signor Sarti's pamphlet, *The Celebrated Florentine Anatomical Venus together with Numerous Smaller Models of Special Interest to Ladies showing the Marvellous Mechanism of the Human Body* depicts a peaceful, recumbent figure, passive and unresisting, and praised as "a perfect model of female beauty" (2) [Fig. 8]. (Although the image on the cover of this pamphlet depicts the Venus fully clothed, the description that follows indicates that this model, like Chovet's, must have been exhibited nude.)

This emphasis on the aesthetically pleasing qualities of the Florentine Venuses is an important point of distinction between the first and second generations of anatomical Venuses, indicative of their repositioning as exhibitory objects within the public sphere, which is, in turn, reflective of both the changed cultural status of anatomy at this time and of a corresponding transformation of the relationship between the practice of anatomy and the anatomical subject. It is significant that Venuses reappear as public exhibits very soon after the establishment of the Anatomy Act in 1832, although Susini's Venuses, at least, were manufactured several decades previously.[23] As we saw in the introduction, the Anatomy Act both reflected and consolidated an important shift in the public reception of anatomy, indicative of its growing acceptance by a bourgeois culture focused on social progress and personal improvement. This shift is clearly evident in the modelling of this

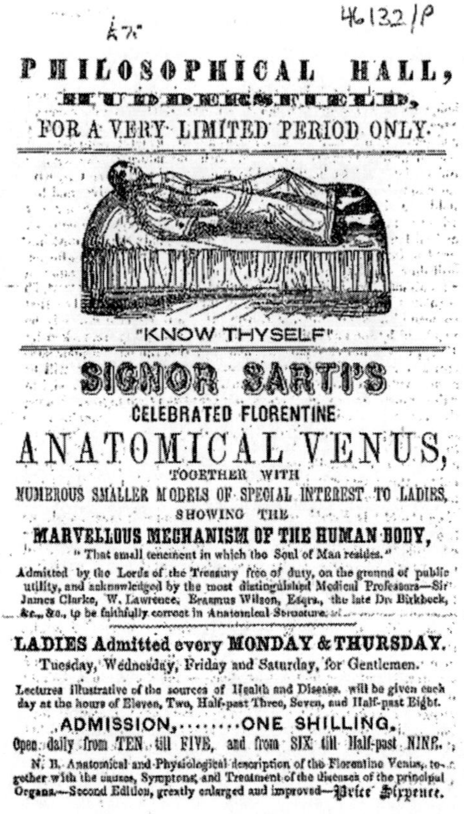

Fig. 8: Cover page of Signor Sarti's *The Celebrated Florentine Anatomical Venus: Together With Numerous Smaller Models of Special Interest to Ladies, Showing the Marvellous Mechanism of the Human Body* (undated). Courtesy of the Wellcome Library.

second generation of Venuses. In the exhibitions of the Florentine and Parisian Venuses, the anatomical subject is no longer represented in the midst of a brutal and punitive act of dissection, is no longer chained and resisting. Rather, these Venuses are peaceful and passive, representing a pregnant body whose encounters with, and opening by, the medical profession are characterised by docility and compliance. In this acquiescence, we see that between the exhibition of Chovet's Venus and that of the Parisian and Florentine Venuses, pain and suffering have disappeared from representations of the pregnant body, marking a significant shift in the contexts and manner in which it could be presented to the public, as well as the significance of its anatomisation.

The disappearance of resistance in the modelling of these Venuses, and the new docility with which they are shown to open themselves to the anatomical gaze, thus reimagines the relationship between anatomy and the body, no longer as one of violent confrontation (as in Chovet's exhibition) but as one of benevolent authority. These new Venuses, with their pearls and silk beds and coiffed hair, also represent a much more affluent and cared-for pregnant body, the middle-class beneficiaries of an anatomical knowledge acquired through the bodies of poorer, more marginal women. As such, this generation of serene Venuses can be seen to represent, aesthetically, the shift from a period of pure anatomical research, in which the knowledge acquired through the dissection of the criminal and the poor had a limited capacity to be applied to living bodies, to one of rapidly improving medical practice, in which this knowledge could be increasingly used to benefit the bodies of the respectable middle classes. We also see this shift articulated in the way that the catalogues for the Florentine and Parisian Venuses address themselves to their audiences.

Unlike grisly spectacles of the sort mounted by Chovet, exhibitions of the second generation of Venuses were advertised as sober entertainments, thus interpellating a very different kind of spectator. The pamphlets for the Florentine and Parisian Venuses take pains to reassure the public that their exhibitions are neither sensationalist nor titillating: "the greatest care [has] been taken to avoid giving the slightest shock to the sensibilities of the most delicate individual," notes *A Description, Anatomical and Physiological, of the Sectional Model of the Human Body, The Parisian Venus* (iv). By emphasising, particularly, their suitability to "the fair portion of creation" (*The Celebrated Florentine Anatomical Venus* 2), these exhibitions are able to broaden their potential audience (and thus increase their profitability), while also consolidating their role as an important part of a respectable bourgeois culture. As Sappol recognises: "The suitability of viewing by women asserted the bourgeois character of the museum" (292), and the catalogues for both the Parisian and Florentine Venuses devote a large portion of their explanatory texts not just to explanations of the actual anatomies on display, but also to reassuring descriptions of the exhibitory space itself. The catalogue for the Parisian Venus advises potential visitors that:

> The exquisite beauty of the Figure—formed on the model of the Venus di Midicis—is exceedingly striking, even at a first glance; but this is nothing to the perfection of workmanship which this section of the anatomy displays. It is a study for all classes; for the young of both sexes, for husbands and fathers, wives and mothers, the healthy and the unhealthy, and especially for the medical student or professor. Nor can we speak too highly of the extreme delicacy with which the whole exhibition (including all necessary explanation) is carried on. Females of the most refined sensibility will find nothing offensive obtruded on

their notice; while knowledge of the most important kind is imparted in a manner of all others most calculated to impress it upon the mind. (iii)

Further reassurances about the delicacy and respectability of these exhibitions were provided by the press reports often included in these catalogues, offering ostensibly impartial accounts of their value. The *Manchester Guardian*, for instance, is cited in the pamphlet for the Parisian Venus, praising it as

> an exquisite work of art, from which we derived much instruction, and have no hesitation recommending it to all who would possess a more definite and correct notion of the delicate organisation of the human frame. Few can see it without finding that, in their general and vague ideas of the human fabric, they have much to unlearn — much more to correct.[...] To the artist, the philosopher, the young medical student; to the young generally, indeed, as a means of physical self-knowledge; to the parent, and especially and above all the mother, this exquisitely constructed model of the human body, internally as well as externally, affords an opportunity for acquiring a knowledge which everyone ought to possess, but which cannot be attained either by drawings, paintings, or other representations on a flat surface, and which no extent of written description can ever convey.[...] We see that ladies are admitted on Tuesdays and Fridays, and we may add, that the model is shown and explained by a medical man, in terms of the greatest perspicacity, propriety, and delicacy; and that there is nothing in the exhibition which need give a moment's pain to any individual. (iv)

This emphasis on the propriety and educational value of these exhibitions — which, as the length of the above quotes indicates, occupied a considerable portion of pamphlets that were usually only four or five pages in length — is highly significant. It should not, however, be understood as a reflection of the dominance of this way of thinking in the 1840s, or as a sign that public exhibitions of anatomical models were uncontroversial to or widely accepted by a popular audience. On the contrary, given that these exhibitions were held within a decade of the public rioting against anatomy schools that preceded the passage of the Anatomy Act in 1832, they need to be recognised as staged in a cultural and historical context in which the practice of anatomy was still highly contentious. If advertising material for these exhibits in the mid-century focused so emphatically on their respectable and educational nature, then, it is precisely because these were not yet established characteristics of these sites, but rather ones in the process of formation. These pamphlets, together with the exhibitions they advertise, play an active role in this formation by insisting on their respectability and encouraging attendance by a respectable female audience.

Hence their curatorial practice tells us about transformations in the cultural status of anatomy at this time and in the spectacular culture developing contemporaneously around it, indicating again the extent to which these were mutually constitutive elements in an emergent public sphere in which the professional and the popular, the medical and the spectacular, were not yet distinct modes of knowledge or representation. In emphasising their own adherence to wider standards of "propriety" and "delicacy," these exhibitions also did much to establish the respectability of anatomy itself. Moreover, they did so in the context of a developing exhibitory culture that was increasingly aligned with official institutions: the first of the World's Fairs would be held in London less than a decade after these exhibitions of the second generation of anatomical Venuses, a period in which new public institutions such as museums and libraries were also developing. Thus the proprietors of mid-century exhibitions of anatomical Venuses, in framing their displays as serious and medical *rather than* sensational and spectacular, serve to draw attention to what was still an emergent and uncertain distinction between these two fields. As a result, however, commercial exhibitions began to occupy an increasingly tenuous position at this time. By the second half of the nineteenth century, the culture of public museums managed by professional organisations was more firmly established, and popular exhibitions like these were increasingly confined to the cultural margins.

We see the consequences of this history in the exhibition of the third and final generation of anatomical Venuses, represented by Spitzner's model, with which we began this chapter [Fig. 3]. With its disembodied surgeons' hands circling its opened torso, brandishing sponges and scalpels, the appeal of Spitzner's Venus for surrealist painters like Delvaux is evident. Spitzner's Venus is a rare example of a surgical Venus, depicting what was then a new procedure: the Caesarean.[24] In contrast to Susini's ecstatic figures, Spitzner's seems decidedly perverse: startled rather than languid, upright rather than recumbent. Where Susini's models recline languorously with arched backs and open mouths, Spitzner's has submissively bound feet. Spitzner's Venus is also the only clothed Venus among the remaining archive of models—which, although we might be tempted to read this as a reflection of changing standards of modesty, serves only to intensify the exposure and penetration of her internal anatomy.

Spitzner's Venus, as part of an itinerant and fairground collection, was also exhibited in a different, and specifically late nineteenth century context, one in which, as Kathryn Hoffman writes, "fairgrounds, museums, morgues, popular photography and private society entertainment—shared techniques of display that merged public knowledge, popular entertainment, death memorial and fairground *flânerie* in fascinating and problematic ways" (140). Fairground Venuses were often hybrid figures that were part medical model, part novelty attraction: the Musée Deranlot, for instance, exhibited a "bare-breasted Venus whose chest rose and fell automatically," while the Musée Spitzner placed in its entrance way a Venus "with a mechanism to simulate respiration" (Py and Vidart 4). Hoffmann notes that such models can be seen as part of another, related kind of popular exhibition: that of "sleeping beauties in the fairground," mechanical or live models of "pretty young women dressed in white satin, with long dark hair," reclining "on silk-draped couches, sometimes in glass cases" (139).

Spitzner's model is one of the few remaining examples of a Venus exhibited primarily in a commercial setting, although it has now been withdrawn definitively from public view.[25] While the Venuses that were produced and circulated originally in professional contexts—such as the Susinis in La Specola in Florence and the Josephinum in Vienna—remain on public display, Venuses like Spitzner's, exhibited in the context of what Vanessa Schwartz refers to as the "lowbrow, almost pornographic, itinerant wax anatomical collections which travelled the fair circuit" (103), have largely disappeared.[26] As the practice of anatomy became a more respectable and established part of medicine, then, and as an anatomised way of seeing the body became progressively more common, popular exhibitions of anatomy became correspondingly more marginal and disreputable. By the time Spitzner's Venus was exhibited at the Foire du Midi de Bruxelles in the early 1930s, their purpose as sites of popularisation for anatomical knowledge about the body had been largely forgotten. What we see if we examine this history more closely, however, is that popular exhibitions of anatomical Venuses have much to tell us about how women's bodies, and their reproductive capacity, were seen through the lens of anatomy, as well as how anatomy saw its relationship to the bodies it dissected, as we will see in the following section.

Embodied Knowledge and Imaginary Anatomies

While anatomy may now disavow its spectacular elements, what anatomical Venuses make so spectacularly evident, and what makes them seem so compellingly weird to contemporary eyes, is the extent to which anatomical science sees the female body through the framework of the cultural signifiers conventionally associated with femininity. Moira Gatens has argued that in addition to our material bodies, we each have an imaginary body, one that is "constructed by: a shared language; the shared psychical significance and privileging of various zones of the body (for example, the mouth, the anus, the genitals); and common institutional practices and discourses (for example, medical, juridical and educational) which act on and through the body" (12). Anatomical Venuses, with their highly aestheticised representations of female reproductive anatomy, might similarly be seen to represent an imaginary anatomy, whose significance is, following Gatens, determined by the language of anatomy, the privilege it accords to various parts of the body, and the institutional practices and discourses of which both professional medicine and public exhibitions are a part. At the same time, Venuses make visible the imaginary of anatomy: how bodies are seen through the lens of anatomy and how anatomy visualises its own relationship to the bodies with which it comes into contact.

What we see in the three generations of anatomical Venuses described in the previous section is three dramatically different interpretations of women's reproductive biology across a period in which anatomy is rapidly repositioning itself and redefining its cultural role: in Chovet's time, anatomy is still a disreputable and controversial practice; the exhibitions of the Florentine and Parisian Venuses coincide with the period in which anatomy is gradually being established as a respectable and important part of medical practice; while Spitzner's model is reflective of a culture in which spectacular displays of anatomy are rapidly losing their claims to a popularising or educational function. This part of the chapter, then, will examine the aesthetics of Venuses as representative of medicine's understanding of the cultural significance of the bodies it depicts, which we can discern by attending to the way that reproductive anatomy is framed and exhibited. As we saw in the previous section, despite the different aesthetics by which they were modelled, all Venuses share the *fact* of their aestheticisation of female anatomy, their insistent framing of the female reproductive organs with apparently extraneous details—the ropes that bind the disciplined body, the pearls of the domestic body, the nightgown of the modest body—which reveal at least as much about contemporaneous cultural constructions of maternity as they do about female reproductive anatomy.

In this, Venuses differ markedly from the models of male, or "standard," anatomy manufactured at this time. While anatomical models of male bodies from this era can also be aesthetically striking—Zumbo's "Anatomical Head" is represented as rotting, for instance, while many of Susini's models are depicted with pained or intensely melancholy expressions—they are not laden with cultural signifiers to anywhere near the same degree as the Venuses. None are ornamented with anything equivalent to their pearls or ribbons. Nor are any clothed, and the manner of their representation does not change so dramatically across time. Alan Bates claims that male equivalents of Venuses, known as Adonises or Samsons, were also produced; however, catalogues of the period published by popular exhibitors, public museums and anatomical modellers make scant reference to any such figures, suggesting they were nowhere nearly so widely manufactured or exhibited as Venuses. Bates himself recognises that "the Adonises and Samsons seem not to have survived, perhaps because when anatomy shows ceased to be commercially viable they were of less interest to collectors" ("Anatomical Venuses" 186). Whereas male figures have traditionally been used to represent "normal" or "standard" anatomy, female figures are primarily gynaecological or obstetric; that is, they depict only those parts of the body the male form cannot be used to show.[27] The female body hence represents the specific; the male body, "the general, the universal, the human" (Grosz 198). Just as contemporary medical imaging has embraced the rhetoric of non-style (as we saw in the introduction to this book), so male anatomical models are represented according to the aesthetics of non-gender. Femininity, then, is not simply a kind of gender, it *is* gender, and more specifically it is its reproductive anatomy.

As Ruth Perry argues, recognisably modern ideas about maternity first emerge during the eighteenth century, constituting motherhood as "a newly established social and sexual identity for women" (205). While these new concepts of motherhood had multiple points of origin and formation, medical discourses can be seen to have played a crucial role in their development and increasing cultural influence. Medicine provided legitimacy and scientific justification for new ideas about the social significance of maternity by claiming to identify the biological sources of these traits in the internal anatomy of women; that is, providing a material and anatomical basis for the emergent cultural ideas about maternity. Gallagher and Laqueur claim that during the first half of the nineteenth century, an increasingly medicalised understanding of sexuality and embodiment led to two important, and contradictory, shifts, both of which are justified in and through anatomical studies of women's reproductive biology. In the first, reproductive anatomy was "increasingly understood" as "the key to women's nature," so that "the essence of Women becomes ever more elaborately sexually embodied" (viii). In the second, however, "women are increasingly conceptualised as people without strong sexual feelings. Whereas it was thought normal for women to be ruled in all their mental states by the activities of their reproductive organs, it was also thought abnormal for them to have pleasurable sexual sensations" (viii).

These interwoven medicalised narratives of femininity, in which female sexuality is increasingly reduced to its reproductive biology just as maternity is progressively detached from sexual pleasure, are embodied in the figure of the anatomical Venus. Both Georges Didi-Huberman and Ludmilla Jordanova have read Susini's Venuses, with their arched backs and ecstatic expressions, as directly representative of medicine's eroticisation of the reproductive female body, a sexualisation of a body increasingly coming to be defined in terms of (but also reduced to) its organs of generation. Maritha Rene Burmeister, however, challenges Jordanova's reading of these Venuses. She cautions:

> Without denying the erotic component to the display of these figures, I would at the same time like to insist that these figures are not simply erotic or even primarily erotic. The meaning of the Venuses was partly constituted by the public manner in which they were advertised and viewed, just as their possible interpretations were shaped by the modellers striving to create a work of art by embodying anatomical knowledge in a female figure from another famous work of art. (51)

The conventions of public display, visual arts and anatomical modelling are hardly opposed—or even easily distinguishable—cultural traditions, however, and the conventionality Burmeister draws attention to here also and very clearly encodes both eroticised images of women, and cultural assumptions about female eroticism. At the same time, the arched backs and opened mouths of Susini's Venuses might be seen not simply as evidence of their sexualisation but also as highly stylised signifiers of their ecstatic acceptance of maternity, their blissful acquiescence to an increasingly sentimentalised view of maternity as the sublime realisation of femininity. That is, Susini's Venuses might also be seen to reflect the emergence of a new middle-class understanding of maternity, and a corresponding reimagination of the reproductive female body, as docile and domesticated, finding new pleasure and fulfilment in its acceptance of a maternal role.

For John Tosh, the anatomical redefinition of women as less sexual in the medical literature of the mid-nineteenth century was directly responsible for a new stereotype, that of the "passionless moral mother" emergent at this time, which Tosh attributes to the increasing role medical discourses played in popular understandings of femininity and female sexuality:

The medical discovery in the 1840s that conception did not require
female orgasm merely reflected assumptions about women's sexuality
which had been current for two generations.[...] The most-quoted
contemporary opinion on this subject is Dr William Acton's in 1857,
that "the majority of women (happily for them) are not very much
troubled with sexual feeling of any kind." [...] [T]he passionless woman
was by now firmly established in respectable middle-class culture. (44)[28]

We see this new understanding of maternity in the modelling of
Spitzner's Venus, whose clothed state replaces the vaginal opening
with a surgical one, and whose expression shows none of the languid
rapture of a Susini figure. Here we see a complete transformation of
the relationship between medicine and the pregnant body described
at the start of this chapter, prior to the coeval development of artificial anatomy and gynaecology: far from being considered a "natural"
event outside the purview of medicine, childbirth as represented by
Spitzner's Venus is a process fully controlled by the disembodied
hands of surgeons, to which the pregnant woman is depicted as a
(rather surprised) bystander.

As popular objects of public display over a period in which a whole
new way of understanding the maternal body—and the modern idea
of maternity itself—is coming into being, part of the cultural function of anatomical Venuses is thus to teach spectators how to see, or
interpret, that body. The public exhibition of anatomical Venuses from
the early 1700s onwards provides an important cultural space in which
this new concept of maternity and motherhood gained authority as the
"truth" of femininity, a truth located in the reproductive anatomy of
women. Their popularity as exhibits over such a long period, enduring well after their function as medical models had been superseded,
attests to their impact in this regard. We see this reflected especially
in the way exhibits of anatomical Venuses were very deliberately and
emphatically didactic: Sarti's exhibition provided five public lectures
a day on "health and disease"; Dr Kahn's Museum of Anatomy had a
regular schedule of anatomical demonstrations; while the Director of
the Musée Quitout also provided lectures on anatomical dissection
using his Venus as the demonstration model.

As this aspect of these exhibitions—their explicit instruction about
and lessons on human anatomy to a lay public—demonstrates, these
spaces were also very specifically and explicitly sites of bodily training.
In a period marked, as Sappol notes, by a "growing consumer demand
for anatomical instruction as part of the curriculum of bourgeois
social identity" (278), anatomical Venuses provided a figure through
which such instruction could be disseminated to a wide audience.

Reflecting a tendency we have already seen in the introduction to this book, these exhibitions not only instructed their viewers how to read the bodies on display (what parts of the anatomy they can see and what biological function these serve) but also how this information should inform their relationship to and treatment of their own bodies, thereby encouraging a compliance with and docile acceptance of the increasing extension of medicine's authority over their bodies and behaviours. This is clearly articulated in the catalogues for the Florentine and Parisian Venuses, which explain in explicit terms how female spectators should understand and care for their own bodies. The catalogue for *The Celebrated Florentine Anatomical Venus*, for instance, directly addresses the importance of the exhibition for female viewers:

> This exhibition presents an UNPARALLELED OPPORTUNITY to Woman of acquiring knowledge *of the greatest personal interest to her*. Every lady, every Mother more especially, ought to avail herself thereof. Let those who from fastidiousness or delicacy shrink from attending the present Exhibition bear in mind that excessive sensibility fosters ignorance, and ignorance opposes the laws of the Creator, the infringement of which alone causes the greater part of that suffering which in the bare contemplation is attended with acute pain to a well-regulated mind. (3)

Exhibitions such as this—the first to specifically target women—have as part of their declared purpose the aim to teach women how to assume the responsibilities of the new maternal subject coming into being at this time, in large part by providing grim warnings about the dire consequences of failing to do so. Women were reminded that they were responsible not only for their own health but for that of their unborn children: "Through ignorance of the delicate being committed to her care ... misery, distress, disease, are daily caused; infant life, to a fearful extent, is sacrificed; and her own constitution often undermined" (3). The contemporary fashion for wearing corsets was an issue of particular concern. The catalogues for both the Florentine and the Parisian Venuses strongly advised against this practice, particularly during pregnancy, referring to the displayed reproductive anatomy of the Venus and her exposed foetus as empirical evidence of its potentially damaging and possibly fatal effects.

Spitzner's museum, which as we saw in the introduction to this book, cautioned viewers to "keep your body clean./And keep it under constant surveillance" (*Grand Musée* n. pag.), also served to remind the spectator of the unpleasant fates likely to befall those who did not. Later exhibition spaces like Spitzner's juxtaposed their centrepiece exhibits of Venuses, representative of an idealised maternal femininity, with cautionary (and often lurid) pathology exhibits of venereal disease, designed to reveal the grisly results of untempered behaviour and unfettered appetites. Such rooms were designed to serve as a salutary lesson about the need for self-regulation and bodily control. Thus, rather like Chovet, fairground exhibitors like Spitzner combined gory sensationalism with claims to be morally uplifting. As Py and Vidart note:

> The directors of anatomical museums adopted the appearance of great popularisers of scientific knowledge and justified the pathology sections of their exhibitions through moralising discourses. In exposing the ravages caused by alcoholism and sexually-transmitted illnesses, a visit to the pathology museum was seen as simultaneously morally-improving and salutary.... It was a matter of capturing the imagination of the public: the hyperrealism of the waxworks, the program of the museum must make the audience tremble with horror and, in order to avoid such a corruption of the flesh, cause the spectator to undertake better social hygiene. (6–9)

These late exhibitions thus made spectacularly evident the flipside of the idea of the responsible self-managing subject of health, the consequences of the undisciplined body and of the dangers it posed for the subject if not properly controlled. Pathology exhibits in these spaces "[b]oth fearfully and seductively suggested that beneath the orderly cognitive surface, the body might be a savage, superficially colonized region of desire, impulse, messy affect, and disease" (Sappol 280).[29] In this way, they reinforced the need to internalise the disciplinary practices required to constitute oneself as a healthy subject, to become the docile subject of modern anatomical practice and to see oneself through its penetrating gaze. That it is through the bodies of women that spectators were encouraged to understand the relationship between anatomy and the body as one of docility and compliance is not coincidental, reinforcing a long association of femininity with passivity while further extending this to all subjects. Anatomical Venuses, as sensationalised or highly aestheticised figures of the naked female form, rendered this newly passive and docile subject of anatomy more palatable to public tastes and cultural standards.

Venuses, with their removable panels, spectacularise the progressive opening of the body to the anatomical gaze, passively offering themselves up to the detached hands of medical practitioners. Thus, Venuses were a key cultural site for the popularisation and dissemination of a new anatomical imaginary, in which the truth of the body is understood as lying hidden inside it.[30] Highly adaptable to the changing mores of their contemporaneous exhibitionary and popular cultures, anatomical Venuses remained so enduringly popular precisely because, as we have seen above, they were significant across a number of different cultural fields, drawing on widely recognisable and familiar visual codes, on the new visual technologies and public spaces emergent at this time, together with the new forms of knowledge enabled by anatomical study, and the new ways of seeing and representing the body this produced.

That models designed originally for medical professionals proved so readily adaptable to the public sphere attests to the consistency of aesthetics and modes of seeing the female body between them, reflecting a close interrelationship between medical knowledge and popular assumptions about female bodies. In this way, the Venuses serve to reveal not only how conventional was this modelling of the body, how mediated through popular thought, but also the extent to which the anatomical gaze came to acquire its current dominance—by simultaneously relying on and disavowing its relationship to the spectacular and popular elements—while entirely reimagining the way bodies were represented and understood. Thus they created subjects that were both passive and self-managing, offering themselves up to the medical gaze while carefully scrutinising their own bodies, constructing them as modern subjects of health perpetually anxious about disease and determined to maximise their vigour and stamina. Hence what these models have to tell us about the history of medicine—the knowledge about the body and reproductive anatomy they convey, where and by whom they were manufactured, the contexts in which they were exhibited—is only part of their significance, as we have seen above. The Venuses also remind us to attend to *how* that knowledge is constructed, providing not just an archive of medical knowledge about the female reproductive body but a striking record of the way that body was culturally valued and understood across the period of their manufacture.

Notes

1 Like many popular anatomical museum proprietors, despite his title, Pierre Spitzner (1834–1896), appears to have had no medical qualification.

2 The most detailed accounts of the Spitzner collection and its history are provided in the collection edited by Blon et al., *Voir Spitzner* (1988), and, more recently, Kathryn Hoffmann's "Sleeping Beauties in the Fairground" (2006).

3 Schwartz locates the original site of Spitzner's museum at the Pavillon de la Ruche at the Place du Château d'Eau (97). Py and Vidart note that there were eight anatomical museums in operation in the Parisian fairgounds of the 1880s, including the Grand Musée pathologique Gross, the Musée Deranlot, the Musée Quitout, and the Musée Groningue. After this time, the number of

fairground anatomical exhibitions began to fall, with three or four in 1911 and then two in 1913 (9).

4 Michelle Bloom argues in *Waxworks: A Cultural Obsession* that the popular wax-works museums of the nineteenth century, such as Madame Tussaud's, grew directly out of the anatomical exhibitions of the early eighteenth century. The first commercial waxwork exhibition was staged by Philippe Curtius in 1766, who originally began as an anatomical modeller (7–8). (See also Vanessa Schwartz's *Spectacular Realities: Early Mass Culture in Fin-de-siècle Paris*.)

5 Georges Didi-Huberman's *Ouvrir Vénus: Nudité, rêve, cruauté* places the anatom-ical Venuses within this longer history.

6 This slippage between anatomical knowledge about the reproductive systems of women and culturally constructed images of female sexuality is reflected not only physically, in the conventionality with which the Venuses are modelled, but also conceptually, in the etymological transformation of the word "Venus" into "venereal" and "venery."

7 See, for instance, Poggesi (11).

8 While full and partial body obstetric models of the female reproductive anat-omy are still produced today, these are not "Venuses" in that they do not conform to the same aesthetic conventions examined in this chapter.

9 While the Wellcome Museum does have an earlier anatomical Venus, dating from the seventeenth century, it is made of wood rather than wax and is much cruder in execution and anatomical detail. I have not come across any documents suggesting this figure was exhibited to a general public in the seventeenth century, nor detailing its use as a teaching model.

10 Partial body obstetric waxworks, showing just the reproductive system and foetuses at various stages of development, were also produced during this period, designed to show foetal positions and complications during childbirth.

11 The partnership between the two ended in a dispute whose nature is not known (de Ceglia 3) but which "involved money and accusations of plagiarism" (Burmeister 31; see also Haviland and Parish 56).

12 Nineteenth-century anatomical modellers, like the French surgeon Louis Auzoux (1797–1880), continued to find a more receptive market among popular museum proprietors than among medical colleges. As Bates notes: "'The dreadful murders committed to procure subjects for dissection' and 'the great opposition shown to the proposal for an anatomy bill' led Auzoux to England in 1832 to demonstrate his new model to anatomy schools, but, while admiring its ingenuity, even those medical men who approved of it thought it no substitute for cadaveric dissection and by 1834 only three had been sold" ("Anatomical Venuses" 184–85).

13 Such courses were a recognised avenue for medical training until the early nineteenth century. It was only after the passage of the Anatomy Act that professionally recognised anatomical instruction would be restricted to medical colleges attached to universities and hospitals.

14 At this time, the relationship between artificial anatomical models and actual bodies was, despite this, quite material: Zumbo's "Anatomical Head" is moulded around an actual skull, while Susini's Venuses, discussed below, were moulded from as many as 200 cadavers.

15 Chovet undertook his medical training in Paris and arrived in London in 1933. According to Miller and Schnalke he moved to Philadelphia in 1774 (although Haviland and Parish give this date as 1770), where he taught for more than a decade at the College of Philadelphia (now the University of Pennsylvania) (Miller 1911; Schnalke 1995). Chovet appears to have relocated

an extensive anatomical collection with him to Philadelphia, which was bought after his death by the Pennsylvania Hospital. In 1884, the entire collection was destroyed by fire.

16 As MacDonald writes, until the Anatomy Act, the only legal source of cadavers for medical colleges was through the penal system: "Until 1832, London's College of Surgeons had been receiving all the bodies of those executed for murder in that city since 1752" (13). (Charitable hospitals were permitted to dissect the bodies of those whose families could not provide for their removal and burial, providing a much greater access to cadavers.)

17 Rackstrow's collection was later sold to the University of Dublin, where at least a part of it has been preserved.

18 Although anatomical exhibitions appear to have been much more popular in the UK than in other European countries at this time, by the mid-1700s Felice Fontana had begun to produce artificial anatomical models for medical instruction and later for public display in La Specola in Florence, while in France Marie Catherine Bihéron (1719–1786) opened her private collection to a fee-paying public every Wednesday (Haviland and Parish 61).

19 Exhibitions were operated by Signor Sarti, Monsieur Esnaut, Madame Hoya and Antonio Serantoni (Alberti 12), as well as, during the following decade, by Dr Kahn and R. W. Reimer, whose larger and more general collections advertised Venuses as their star attractions.

20 As Samuel Alberti recognises in his article on Joseph Towne, the English anatomical modeller, Britain lagged far behind Italy and France in its expertise in the production of artificial anatomy during this period (8).

21 The Museo delle Cere Anatomiche in Cagliari contains only partial-body models.

22 As Katherine Park has shown in *Secrets of Women*, the history of autopsies in Italy, at least, complicates the usual account of its historical development, as these were often practised in domestic spaces, such as post-mortems on aristocratic women who had died in childbirth. However, as Park demonstrates, these practices were often undertaken for religious or familial reasons, rather than medical ones, and thus occur in a different cultural space than that of the anatomical tradition that arose within the context of modern medicine from the 1600s.

23 Precise dates for some of these exhibitions are a matter of conjecture. For instance, although the date of exhibition for the Florentine Venus is not recorded in its catalogue, the pamphlet for the Parisian Venus, which toured the UK in 1844, refers to the Florentine Venus in a way that suggests this latter was first exhibited just before the Parisian model. The pamphlet cites a commendation by The *Birmingham Advertiser*, which begins: "Having twice paid a visit to the Florentine Venus, our curiosity was no so great as it otherwise would have been, and it was not till yesterday we paid a visit to this wonder of art" (Alison iii). The two Venuses also seem to have been very similar in style and execution, with the notable exception that the Parisian Venus appears to have been modelled to show the entirety of the anatomy, not just that of the reproductive system: the pamphlet describes the anatomy of the face, neck and limbs in detail.

24 Before the invention of anaesthetics, writes MacDonald "[d]eeper operations were only performed in extreme emergencies, in the knowledge the patient would probably die" (28). The first successful Caesareans were performed during the mid-1800s, but were still far from commonplace when this Venus is likely to have been produced. Hoffmann provides descriptions of two more

surgical Venuses from the Spitzner collection, both of instrument-assisted births, although these figures appear to have been subsequently lost.

25 The collection was discovered in a storage facility in the late 1970s, before being offered for sale in the mid-1980s. The auction catalogue (*Grand Musée Anatomique et Ethnologique du Dr P. Spitzner*) provides a record of what remained of the collection at this time. Much of Spitzner's collection (including the Venus reproduced here) was bought by the Musée d'Anatomie Delmas-Orfila-Rouvière (run by the University of Paris's Anatomy Department), which has also recently closed. Spitzner's Venus is now in permanent storage.

26 One reason for this is that many of these pieces were destroyed during the crackdown on popular anatomy museums by vice squads in the mid- to late nineteenth century, described in the previous chapter.

27 The collection of eight standing écorché figures at the Josephinum in Vienna does include two whole-body female models (not Venuses), although this appears to be one of the very few exceptions to this rule.

28 See also William Acton, *Functions and Disorders of the Reproductive Organs* (1865).

29 Although the rooms dedicated to venereal disease exhibits in the fairground museums were clearly highly spectacularised, it is important to remember that professional medical modellers such as Jules Talrich and Louis Auzoux also produced cautionary moulages (block-mounted partial-body figures), in order to demonstrate the dangers of alcoholism, masturbation and promiscuity, alongside their standard anatomy and pathology models, indicative of the link between moral degeneracy and anatomical pathology that medical discourses reinforced at this time.

30 Katherine Park has argued that women's bodies are represented in medicine as hiding "secrets" from the medical gaze, especially regarding "information relating to generation and the female genitals" (91).

CHAPTER 2

Lost Manhood: Turn-of-the-Century "Museums of Anatomy" and the Spermatorrhoea Epidemic

By the first decades of the twentieth century, the term "museum of anatomy" was no longer associated with respectable and improving exhibitions that claimed to educate the public about bodily health and care. Whereas the exhibitions of Venuses established their place in an emergent bourgeois public sphere by emphasising their suitability for, and importance to, female spectators, by the turn of the nineteenth century popular anatomical museums were exclusively "for men only," and were to be found in transitional, transient locations around entertainment districts, train stations and docks. Featuring lurid displays of sexual anatomy, especially diseased sexual anatomy, the purpose of these exhibitions was primarily to serve as advertising for their associated "medical institutes." Although very little documentation of these museums has survived, Brooks McNamara provides a detailed description of their curatorial style:

> The front windows of most medical museums were designed to stop the idle passer-by in his tracks with some kind of arresting tableau.[…] As a tourist entered the museum, under a sign reading "For Men Only," he found himself in a kind of antechamber, usually nothing more than a corridor lined with glass cases. In the corridor the proprietors displayed wax or papier-mâché models of the earliest and least loathsome stages of venereal disease, voluptuous female anatomical models, a few live monkeys, snakes or birds, or a miscellaneous collection of curiosities like those found in most dime museums.[…] Once inside the main room, the assault on the patron's nerves began in earnest. Everywhere about him were glass cases filled with hideously diseased organs modelled in death-like wax or luridly painted papier-mâché. The lights were low, the atmosphere hushed and funereal. Case after case displayed gaping sores and hideous deformities attributed to syphilis, gonorrhea, or to that nameless terror of the nineteenth century, the "secret vice," masturbation. Finally, in the darkest and most remote corner of the

museum, the already shaken tourist came upon the horror of horrors, an innocuous glass case, a light flashed on and he was confronted with the ghastly visage of a smirking a drooling idiot boy and the awful legend "Lost Manhood." At this point a solicitous "floor man" would appear from nowhere and begin to talk to the frightened patron. If it appeared that the customer was in need of medical aid—or could be convinced that he was—he was steered upstairs to the "medical institute" run by an "eminent specialist" in the various secret diseases.[...] Once inside and in the clutches of the "specialist," it was unlikely that the customer would emerge again until he had spent at least five dollars and often as much as twenty dollars on worthless medication or treatment. (39–40)

As spaces "for men only," anatomical museums of this period made no claims to educate their audiences, instead repositioning themselves—and thus redefining the anatomy museum—within the context of a rapidly expanding commercial sphere in which manufacturers of patent medicines took full advantage of the development of new visual technologies and forms of popular media to market their wares to a large audience. At the same time, these spaces are nonetheless and recognisably part of the same exhibitory and disciplinary culture as earlier museums.

We see this in the way that their lurid exhibitions attribute "lost manhood" to excessive sexual practices such as masturbation or promiscuity, for which the sufferer could be held personally responsible, and whose condition was seen to have been brought about by a lack of self-discipline or control. Such failures would not only exclude one from polite society but would also be clearly visible on the body itself, in the "ghastly visage of a smirking a drooling idiot boy" that makes spectacularly manifest on the surface of the body the diseased anatomy of its interiority. In this way, these new museums, dubious and marginal as they seem, were nonetheless reinforcing the disciplinary function of early anatomical museums, and their belief that the health and condition of one's body was a reflection of the care taken of it.

Thus, while the "museums of anatomy" that flourished in the second half of the nineteenth century developed directly from the earlier tradition of commercial anatomical exhibitions described in the previous chapter, they also represent its institutional reinvention and the transformation of its cultural significance: they stage a very different form of public spectacle (focusing on pathology rather than "normal" anatomy exhibits), for a newly restricted "public" (limited to men only), in a geographically more marginal position (on the city's fringes or transient zones rather than its centre), using markedly different exhibitory techniques and technologies (drawing heavily on new forms of mass marketing). Despite these differences, the cultural work performed by the late nineteenth century of public museums of anatomy remained, in many ways, consistent with that of previous generations. Just as the modelling and exhibition of anatomical Venuses examined in the previous chapter reflected wider cultural assumptions and understandings about femininity and maternity, so does the spectacularisation of "lost manhood" have much to tell us about late nineteenth-century conceptualisations of masculinity and male sexuality. As we have seen in the previous chapter, particular bodies become the object of public fascination and medical scrutiny during those periods in which their cultural significance and status is undergoing rapid change. At the end of the nineteenth century, the body around which this attention accrued was the sexual male body. This is especially noteworthy because, historically, the male body has almost always been positioned as the subject of spectacle (or the gaze), rather than its object. This makes public anatomical museums of this period a valuable source of information about how male sexuality was seen and understood at this time.

This chapter will thus consider how the generation of anatomical museums that flourished subsequent to Kahn's museum in the late nineteenth and early twentieth centuries represented both a transformation of earlier styles of public anatomical exhibition and an intensification of their rhetorical focus on the importance of self-cultivation and self-discipline for the maintenance of good health. It will begin by tracing the development and treatment of "lost manhood" in museums of anatomy during the last decades of the nineteenth century and first decades of the twentieth century, positioning these both within the history of medicine, as part of a long tradition of writing on "seminal loss," and within their commercial context, as part of the exhibitory and advertising cultures of their time. The second part of the chapter will focus on the translation of discourses about "lost manhood" into medical terminology as the new condition "spermatorrhoea." Considering its treatment in both licensed and quack medical contexts, this section of the chapter will examining its (re)imagining of male fluidity and what this tells us about the rapid changes in dominant understandings about masculinity taking place at this time and the role of spectacularisation within it.

Medicine and Male Body Fluids: The Anatomy of Seminal Loss

Although the museums of anatomy "for men only" that flourished in the second half of the nineteenth century did occupy a much more marginal and precarious cultural position than previous generations of anatomical museums, they nonetheless remained an important part of the institutional and discursive networks by which changing ideas about masculinity circulated, and an influential site of production of both public images and popular medical discourses about male bodies and sexualities. The transformation of the earlier style of anatomical museum into a commercial site for the trade in (and construction of) "lost manhood" begins, as we saw in the introduction to this book, with the collaboration between Kahn and the even more dubiously credentialled "Dr" Jordan, a distributor of patent tonics for men. After the closure of Kahn's museum, as detailed in the introduction, Jordan went on to open anatomical museums in the UK, Australia and the USA, under a variety of pseudonyms. In the process, he established a set of business and curatorial practices that would provide the template for a whole new generation of anatomical museums. It is thus to Jordan, almost single-handedly, that we can attribute the transformation of the popular anatomical museum into a quack medical institute and marketing centre in the last decades of the nineteenth century. This shift is evident in the publications these sites produced. Catalogues for anatomical Venuses were short descriptive texts designed to advertise their exhibition and explain the models' anatomy and function, but the proprietors of later museums, like Jordan and Kahn, extended these documents into handbooks of considerable length (sometimes of 100 pages or more), which also incorporated (pseudo-)medical treatises on male sexual health. Clinical as much as exhibitory, anatomical museums after Kahn's represent a profound transformation in the nature and purpose of these museums, translating the spectacular style of earlier museums into the rhetoric of their handbooks and other published material, (re)producing a new relationship between the spectacular and the medical—and the view of the body that this in turn reflects—emergent in the late 1800s.

It is through the publication details of these handbooks that we are able to recover the institutional history of Jordan's museums and to track the cultural work that spectacularised displays of male sexual anatomy played in the discursive reconfiguration of medical and commercial cultures in the late nineteenth century. After dissolving his collaboration with Kahn, Jordan opened the London Anatomical Museum, whose proprietor is identified on the cover of *The Illustrated and Descriptive Catalogue of the Subjects contained in the London Anatomical Museum to which is annexed the Guide to Masculine Vigour* as Robert Jacob Jordan. This museum closed in 1863, at which time Jordan relocated to San Francisco, where, as Louis Jordan, he ran the Pacific Museum of Anatomy and Natural Science, publishing *The Handbook of the Pacific Museum of Anatomy and*

Natural Science, which also contained *The Philosophy of Marriage: Being Four Important Lectures on the Functions and Disorders of the Nervous System and Reproductive Organs* in 1865. By 1867, Jordan was in Australia, where, as Henry Jacob Jordan, he opened Dr Jordan and Dr Beck's Anthropological Museum in Melbourne. He was still there in 1869, when he became embroiled in a court case after bringing a libel action against *The Age* newspaper, which had run a series of condemnatory articles calling for his museum to be closed.[1] During this trial, Jordan's business partner, Dr Beck, was exposed as a figment of Jordan's imagination. Despite being publicly ridiculed for this in the Australian press, Jordan was not deterred from inventing a new collaborator for his next museum in Philadelphia: the cover of *Practical Observations on Nervous Debility and Physical Exhaustion: to Which is Added an Essay on Marriage, with Important Chapters on Disorders of the Reproductive Organs. Being a Synopsis of Lectures Delivered at the Museum of Anatomy*, published in 1871, lists a Dr Davieson as co-author [Fig. 9]. The end of the 1870s saw Jordan in New York, where as Henry Jordan he ran the Museum of Anatomy in 1878, as Philip J. Jordan the New York Museum of Anatomy in 1879 and as Louis J. Jordan the Museum of Anatomy, Science and Art in 1881.

As the brief account above shows, Jordan routinely included essays on male sexual diseases and dysfunctions as appendices to his museum handbooks, and these texts allow us to trace the way in which he reinvented the popular anatomical museum as primarily for the purpose of advertising his much more profitable trade in patent tonics for "lost manhood." In *Practical Observations on Nervous Debility and Physical Exhaustion*, he advises his readers:

> The fearfully abused powers of the human generative system require the most cautious treatment: our studies for many years past have been exclusively directed to the treatment of the debility and diseases resulting from self-pollution, venereal infection, loss of sexual power, and such complaints as arise from disorganisation of the reproductive powers, whether constitutional or acquired; and in conclusion we may observe that all who apply for advice or assistance may always depend on that inviolable secrecy, sympathy, and skilful attention which have already proved the basis of the most extensive practice in special diseases in the United States of America. (Jordan and Davieson x)

Jordan assured his readers that his treatments could be mailed in discreetly unidentifiable packaging, sent to anonymous or pseudonymous recipients, and that a range of flexible payment plans was available on request.

Jordan's use of what were ostensibly museum handbooks to advertise a mail order patent medicine business is a reflection of the close

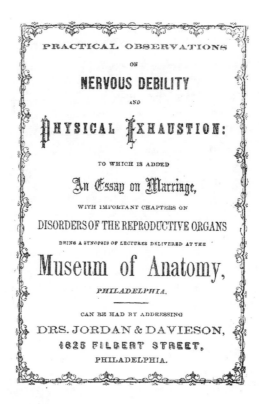

Fig. 9: Cover page of Jordan and Davieson's *Practical Observations on Nervous Debility and Physical Exhaustion* (1871)

but constantly transforming relationship between the exhibitory methods used in public anatomical museums, those of mass marketing used in a burgeoning commercial sphere, and those of popular medical discourses in the years before genito-urinary specialisations were an established branch of medicine. Popular "museums of anatomy" thus need to be understood within the history of advertising—and more particularly the flourishing market in patent medicines for male sexual problems—as well as medical discourses on male sexual health. It is at this new and very productive point of intersection between popular medical discourses, public anatomical exhibitions and mass marketing that the nineteenth-century anatomical museum flourished and its focus on "lost manhood" developed.

As we have already seen in the introduction to this book, patent medicine companies were among the first and heaviest users of mass-marketing technologies, taking full advantage of new visual technologies and media for circulating information to a mass audience, including illustrated flyers and mail circulars, newspaper advertisements, photography and film, handbooks and brochures. Jordan's own uptake of such technologies—further exploited by the generation of anatomical museums "for men only" that would follow his own—thus served to reposition such museums culturally, not simply by making them progressively more disreputable but also by expanding their potential clientele dramatically, circulating information about their treatments to a much wider audience than museum visitors. While these museums were undoubtedly much sleazier and more marginal than their forebears, then, their cultural impact was nonetheless exponentially greater.

Although Jordan's own museums were clearly very ephemeral and run on a small scale, the style of exhibition he developed gave rise to a much larger chain of museums in the following decade. In the early twentieth century, the Reinhardt brothers—twins Wallace and Willis and brother Frank—established what may have been the largest chain of anatomical museums in the USA. The Reinhardts established a series of medical institutes—including the Wisconsin Medical Institute, the Heidelberg Institute, the Leipzig Doctors and the German-American Doctors—each of which operated several museums. Stewart Holbrook calculates that, at the turn of the century, the brothers owned a controlling interest in between 30 and 40 anatomical museums across the Midwest, and were grossing $100,000 per annum from the sale of their tonics (76–84). Although only Wallace Reinhardt was licensed to practise medicine, and even his licence was revoked by the Minnesota Board of Medical Examiners in 1900, the Reinhardts were never successfully prosecuted—partly because the patent medicine industry would only start to be regulated in the USA after the establishment of the Federal Drug Authority in 1907, and partly because unlicensed practice of medicine was itself not illegal in the USA at this time.[2] As Holbrook notes, by the start of the twentieth century, these businesses were beginning to bifurcate, as

> the Lost-Manhood and the Secret-Diseases-of-Men racket had grown into organised big business. By the turn of the century it was progressing along two general lines. There were the mail-order operators who diagnosed, treated, and cured sufferers by way of post-office or express; and the museum operators who might also use the mails but who depended chiefly on the walk-in trade at the "free anatomical educational museums" they maintained in the Skid-row or Bowery-type districts of towns and cities. (71)

Such companies clearly found a large and receptive market: the Dr. Rainey Medicine Company, whose "Lymph Compounds" were

Fig. 10: Part of the public sexual health campaign by the Social Hygiene Division of the War Department. Lantern slide (c. 1915). Courtesy of the National Museum of Health and Medicine.

sold as "the Greatest Known Treatment for Weak Men," employed 30 girls for their mail order business in 1910. Circulars and handbills for "lost manhood" tonics were so ubiquitous that one of the first public education campaigns about sexually transmitted diseases (initiated by the American Social Hygiene Association during the First World War) warned specifically against them [Fig. 10]. They were targeted again in the early 1920s, in the poster series the American Social Hygiene Association developed for civilians from its wartime campaign [Fig. 11].

Thus, although this generation of anatomical museums "for men only" occupied a much more marginal cultural position than that of the previous generation, and made no claims to have an improving or educational function, their cultural and commercial influence undoubtedly reached much further. These museums both benefited from and contributed to the close relationship between popular medical discourses, public anatomical museums and mass-marketing operating at this time, at whose intersection cultural anxieties about masculinity coalesced. We see one instance of this close relationship in the way the advertising rhetoric of popular texts (such as museum handbooks and commercial advertising) largely reproduced what were then mainstream medical attitudes towards male sexuality, which in turn lent these a certain authority and no doubt contributed to their commercial success. The 1920s fad for "rejuvenation" therapies is a case in point. The best-known commercial practitioner of such treatments was John Brinkley, who began as a "floor man" in turn-of-the-century anatomical museums and later made his fortune by opening a series of clinics specialising in virility treatments. Brinkley's practice was to transplant

Fig. 11: "Ignore Fake Advertisements," part of the "Keeping Fit" public sexual health campaign by the American Social Hygiene Association. Poster (1919). Courtesy of the Social Welfare History Archives, University of Minnesota Libraries.

goat glands into human testicles. He never opened a museum of his own but instead advertised his clinics extensively in the popular press, and his innovations in advertising are considered his most important and influential legacy: while Brinkley used print advertising like circulars to advertise his work, he also bought a radio station on which he hosted a popular call-in consultation show, "Medical Question Box." In the earliest days of radio, Brinkley established what would later become the standard format for commercial radio, positioning medical marketing at its very centre, and selling vast quantities of his patented tonics to the public.[3] In this way, although Brinkley was in many ways a marginal figure within the medical culture of his time, which tried repeatedly but without success to curtail his business operations, he was also very much a product of his era, embodying its contradictions while enjoying an enormous popular influence: he was a dangerous quack with actual medical qualifications; his treatments sound preposterous and yet were consistent with those of prominent and respected medical figures; he was pursued (mostly unsuccessfully) by the American Medical Association for decades, but for his aggressive marketing methods rather than his lamentable tendency to maim or kill his patients.

Although Brinkley was indisputably a quack, the principle underlying his practice—that "rejuvenation" of the male sexual anatomy would restore not only men's virility but their overall strength and manliness—was supported in and by more orthodox medical research of the day.[4] Research into the implantation of animal glands into human testicles, to improve virility and rejuvenate the general constitution, was also being undertaken by the Russian physician Dr Serge Voronoff and the American medical professor Dr Frank Lydston at this time, both of whom were also experimenting with testicle transplants. The experimental biologist Eugen Steinach advocated transplanting testes from heterosexual men onto the bodies of male homosexuals. Of one experimental subject, who identified as homosexual, Steinach recorded that twelve days after "the 'surgical exchange,' the homosexual subject reported having erections and erotic dreams of a heterosexual nature. He had sex with a female prostitute six weeks after surgery and many times subsequently. His voice became deeper and his body more masculine. In less than a year he married" (Sengoopta 467). Steinach also performed vasectomies as a rejuvenation cure, and surgeries of this kind, designed to virilify the effeminate or impotent, remained in medical vogue for a considerable period afterwards.

It is precisely because virility was so widely constructed as failing and problematic by professional medical figures at the end of the nineteenth century and the beginning of the twentieth century that the idea of "lost manhood" peddled by popular anatomical museums, and their wider discursive construction of masculinity as debilitated, was able to gain so much cultural traction. Their spectacularisation of a diseased and pathological male sexuality was reinforced by, and entirely consistent with, concerns not only about virility but also, more generally, about (sexual) degeneration and degeneracy that underpinned much public health and medical discourse at this time. In the last half of the nineteenth century, men were instructed by a wide variety of sources that they had many reasons to worry about their bodies and sexuality. The purity and hygiene movements of this period—a loose collective of community organisations calling for reform and public education mostly on issues related to prostitution and sexually transmitted diseases—played an important role in this process.[5] As Robert Darby notes, these organisations played a pivotal role in moving the focus of public attention from female bodies and sexualities to male, by "shifting blame for prostitution and venereal disease from female depravity and economic need to male sexual desire" (83). Whereas earlier public health movements had focused on the problem of female sexuality, Darby argues, male sexuality was increasingly viewed as "*the* sexual problem of the nineteenth century, responsible not only for prostitution (called into existence by the demands of male lust) and thus venereal disease, but all the additional health problems supposedly caused by masturbation and sexual excess" (83; original emphasis).

When male sexuality was not represented as a source of social problems, it was represented as itself problematised by the social conditions of contemporary life. Whether in medical texts such as George Miller Beard's *American Nervousness* (1881) and *Sexual Neurasthenia* (1884), which documented an epidemic of exhaustion and lack of "nerve force" among American men, or in the work of the physical culture exponent Bernarr Macfadden, editor of the journal *Physical Culture*, and author of the popular guide *The Virile Powers of Superb Manhood: How Developed, How Lost, How Regained* (1900), men were advised that their "natural" vigour and virility were being endangered by the artificial conditions of modern urban life. Excessive reliance on modern comforts was particularly to blame, and had rendered men, en masse, soft and unmanly, "civilised" to the point of effeminacy. Macfadden's advice to men was to toughen themselves up through a rigorous disciplining of their bodies, eschewing sedentary habits for daily exercise, warm baths for cold plunges, a rich diet for light meals, artificial heating for open windows, etc. The "first duty of every male human adult is to be a man," he explains at the start of *The Virile Powers of Superb Manhood*, for

> if you do not possess this virile manhood your imperative duty is to strive for its acquirement, even if necessary for the time being to sacrifice every other purpose in life. For if you are not a man, you are nothing but a nonentity! A cipher! And as long as you remain in this emasculated condition, your powers and capacities in every way will be bound by your weakened condition. (6)

Championing the self-cultivation of bodily strength as a means by which to counter the over-civilising effects of contemporary urban life, Macfadden encouraged a belief in bodily strength and health as perfectible, but as constantly undermined by unnatural habits that debilitate masculinity:

> Ignorance of the facts in reference to the sexual instinct that should be as plain as the noonday sun to every human being, this, together with

lack of knowledge of the great laws of health, so necessary in order to build vigour and symmetry of the body, have resulted in filling civilised countries with a host of pygmy men. Immediately after birth they come in contact with abnormal influences. They are encumbered with clothing that discourages rather than encourages muscular movements; they are compelled to breathe foul air when the weather is cold; they are always overfed; the bottle often does duty for the female breast, and they come in contact with all sorts of conditions that tend to depreciate vitality. Of course over half are killed by all this, and those that survive are greatly weakened, and never attain the superb manhood that should be their inalienable right. (16–17)

What modern men were lacking, according to Macfadden, was proper bodily discipline. It was only by correctly disciplining his body — controlling its appetites, resisting the easy comforts of domestic urban life, hardening the body by cultivating its musculature — that one's virility and manliness could be protected. The photographs of posing muscular bodies Macfadden included in his *Physical Culture* journal were as spectacularised as those of the diseased bodies found in the popular anatomical museums of this period, and were used to make a similar point: the cultivation of a healthy body required an attentiveness to practices of bodily training and the exercise of self-discipline.

While the spectacularisation of "lost manhood" found in the anatomical museums of the late nineteenth and early twentieth centuries hence profitably exploited widespread concerns about masculinity and male sexuality that were specific to the nineteenth century, it also drew heavily on a much longer tradition of popular medical writing on the consequences of "seminal loss" for the male body. In order to fully understand the significance of the "lost manhood" trade centred in popular anatomical museums in the late nineteenth century, it is thus necessary to understand the history of medical writing on seminal fluidity and its "loss" on which these drew. From classical Greek and Roman medical literature onwards, ejaculation and seminal loss have been the subject of enduring and remarkably consistent medical concerns; they were believed to have a potentially deleterious effect on male vigour and virility by depleting the male body's "vital heat." The eighteenth-century physician Samuel Auguste Tissot cited a wide range of these in his *Onanism: or, a Treatise Upon the Disorders Produced by Masturbation: or, the Dangerous Effects of Secret and Excessive Venery*: semen is "extracted from the head" (Hippocrates, quoted 48), a "portion of the brain" (Akmaeon, quoted 50), and a "running of the spinal marrow" (Plato, quoted 51).[6] The extent to which seminal loss has been constituted as a medical problem is reflected in the original meaning of the term gonorrhoea, which was used to refer generally to "flow of seed," rather than to a specific sexually transmitted disease, a "confusing double usage [that] persisted until the eighteenth century," as Darby notes ("Pathologising Male Sexuality," 286 n. 13).

Although the invention of the term "gonorrhoea" is usually attributed to Galen, it was Aretaeus who wrote the most detailed early account of this condition, which is striking for how strongly it foreshadows and shapes nineteenth-century writing on "lost manhood" (as well as its medical equivalents, such as "seminal weakness" and "sexual debility," on which this drew). Aretaeus's description is worth quoting at length, as it came to define medical writing on this topic for close to two millennia. For the patient afflicted with gonorrhoea, Aretaeus writes:

> the seed flows, nor is it possible to suppress it even in the time of sleep, for whether one sleeps or is awake its flow is incessant, and imperceptible [...] [It] is moist, thin, cold, without colour and does not answer the purposes of generation.[...] [I]f young men suffer from the disease, their whole habit is changed and they feel the consequences of age, a resolution of the nervous system takes place, the patients are sluggish, lifeless, torpid, dull, weak, curved, inactive, pale, white, effeminate, have an aversion to food and are frigid, a heaviness of the members of the body, with numbness of the legs, takes place, they are remiss and languid in all their actions: this disease frequently lays the foundation of *paralysis* or a resolution of the whole nervous system, for how can it be that the nervous energy should not suffer, when nature so powerful in generating life is rendered frigid and cold? Seed from its vivifying quality makes us men, and imparts heat, agility, activity, roughness, a manly voice, and courage, it likewise renders us fit to perform all the operations of both mind and body, a proof of which men themselves exhibit: on the other hand, those who are not possessed of this vivifying power are weak, full of wrinkles, have a shrill voice, are without hair, beardless and effeminate, of which eunuchs are a striking proof. If any man is not profuse in lavishing his seed, he becomes strong, courageous and bold, nor is he afraid to encounter even wild animals; the prudent and temperate among the wrestlers give testimony to this assertion, for those who by nature excel others, from their intemperance frequently become weaker than persons who were naturally endowed with an inferior degree of strength; while such as are naturally inferior excel their superiors, which arises entirely from their abstinence and temperate mode of life, nor is animal strength generated from any other thing than seed; its vivifying power therefore contributes much to health, strength, fortitude and procreation. (224–27)

Like the nineteenth-century writers on "lost manhood," Aretaeus associates "seminal loss" not only with the physical depletion of virility but also with failures of mental or emotional self-control. We see this in the way Aretaeus's account proceeds from an anatomical description—with its central, localised symptom of an incessant and imperceptible flow of semen—to one of general constitution and deportment. The reason men need to be so careful about the loss of semen, as Aretaeus writes here, is because this is the source and essence of manhood: it is the "[s]eed [...] that makes us men, and imparts heat, agility, activity, roughness, a manly voice, and courage, it likewise renders us fit to perform all the operations of both mind and body."[7] Ejaculation is potentially dangerous because it depletes men's vigour and heat, makes them "shrill" and effeminate, marked by a tendency to become "lifeless, torpid, dull, weak, curved, inactive," to be mentally as well as physically soft, sluggish and weak.[8]

The ongoing influence of Aretaeus's views on seminal loss on subsequent medical literature is evident in the very long history in which it continued to be reiterated, in almost identical terms, in both professional and popular medical texts. John Hunter, writing fifteen hundred years after Aretaeus, described "seminal weakness" as a form of incontinent ejaculation in a passage that closely recalls that of Aretaeus—not least in his framing of this as a form of venereal disease. "Seminal weakness," in Hunter's account, is

> a secretion and emission of the semen without erection.[...] The secretion of the semen will be so quick that simple thought, or even toying shall make it flow.[...] The spasms upon the evacuation of semen in such cases are extremely slight, and a repetition of them soon takes place; the first emission not preventing a second; the constitution being all the time little affected. When the testicles act alone, without the accessory parts taking up the necessary and natural consequent action, it is a still more melancholy disease; for the secretion arises from no visible or sensible cause, and does not give any visible or sensible effect, but runs off similar to involuntary stools, or urine. It has been observed that the semen is more fluid than natural in some of these cases. (206–207)

Hence, these concerns about "seminal loss," and the "weakness" in the male reproductive anatomy this is likely to cause were widely accepted as medical fact over an extremely long period of time. It was precisely because seminal loss was so widely and unquestioningly seen to diminish masculinity that the later commercial trade in "lost manhood" was able to flourish; although spearheaded by quacks and spectacularised in sleazy anatomical dime museums, it nonetheless was able to draw on the authority of fifteen years of mainstream medical writing in support.

While Hunter's own text, like others written primarily for and by medical specialists, would have been read by a relatively small professional community, his understanding of "seminal weakness" emerged within and was shaped by the much broader field of popular writing on men's sexual health also proliferating at this time, much of which was published in the commercial sphere. Indeed, Robert Darby notes that Hunter was the first to use the term "seminal weakness" in a professional medical context, having imported this term from the commercial sphere in which it was then widely used. Advertisements for patent tonics by empirics and their distributors were featured in broadsheet newspapers dating back to at least the start of the eighteenth century, and thus constitute a direct precursor to the nineteenth-century museum hand-books on sexual health. Accordingly, Hunter's writing, and professional medical literature more generally, is part of a surrounding commercial culture on which it both drew and further reinforced.

In the same decade in which Hunter was writing *A Treatise on the Venereal Disease* and undertaking his anatomical research, one of the most prominent and wealthy men in England was the medical entre-preneur Dr Samuel Solomon, whose famous *"Cordial Balm of Gilead"* was advertised extensively in the popular press, as was his *A Guide to Health; or, Advice to Both Sexes. With an Essay on a Certain Disease, Seminal Weakness, and a Destructive Habit of a Private Nature*. In this text, Solomon provides his own account of the dangers of seminal loss for the male body. He asserts:

> The immoderate evacuation of semen is not only prejudicial on account of the loss of that most useful humour, but likewise by the too frequent repetition of the convulsive motion by which it is discharged; for the highest pleasure is followed by a universal revulsion of the natural powers, which cannot frequently take place without enervation. Besides, the more the strainers of the body are drained, the more humours they draw to them from the other parts, and the juices being thus conveyed to the genitals, the other parts are impoverished. Hence it is, that, from excessive venery, all the various symptoms of lassitude and debility ensue, and are increased by a perpetual itch for pleasure, which the mind contracts as well as the body, and from whence it follows that obscene dreams, frequent erections, and involuntary emissions, bring the flower of youth to premature old age. (33)

Although Solomon's account of seminal losses draws on a humoral theory of the body that was already outmoded in the late eighteenth century, he also rearticulates the deeply entrenched medical assump-tion, traceable back to Aretaeus, that "immoderate" sexual activity exhausts the constitution as a whole.

If Solomon was able to draw so successfully on established medical opinion and to translate this so profitably into a commercial context, it was because both his business practices and his advertising rhetoric had a strong precedent stretching back to the start of the eighteenth century. That is, Solomon drew on a pre-existing commercial culture, as well as medical authority. During the same period in which anatomical Venuses were first being exhibited to a general public, distributors of tonics for "sexual debility" and "seminal weakness" were also beginning to mass-market their wares through newspaper advertisements. A single issue of *The Universal Spectator and Weekly Journal*—a publication of only four pages—from 1733, the same year that Chovet's anatomical Venus first went on public display, includes an entire page of advertisements for such tonics, which included a "transcendent balsamick restrictive electuary," a tonic for "the Venereal Distemper," and another for "Dr Nelson's most wonderful panacea for confirmed Venereal Lues." All reflect the same combination of commercialisation and medicalisation seen in Solomon's advertising. The advertisement for Dr Nelson's panacea exemplifies this, describing gleets and "seminal weakness" as

> The Bane of Virility, or Manhood, in the one Sex, and the Destroyer of Fertility, or the Bearing of Children, in the other, whether from ill-cur'd Venereal Infections (than which nothing is more common) or from inordinate Coition, or Self-Pollution (that cursed School Wickedness, which spoils all our Youth, by nipping their Manhood in the Bud) or from involuntary Emissions at Night in the Sleep, or in the Daytime upon Stool, or with the Urine, or from Falls, Blows, Strains, Wrenches, or the like, which drain and dry up the Seminals, and wither, as it were, the Generative Faculties, causing Impotency in Men, the Fluor Albus, and Barrenness in Women, and in the long-run, Melancholy, Vapours, Decays of Nature, and Consumptions. (*Universal Spectator* "Dr Nelson" n. pag.)

Although written for a popular, non-specialist audience in an unregulated market, in which the modern practice of medicine had not yet been established, these advertisements nonetheless reflect the growing reliance on medical discourses and professional titles as markers of authority and proof of efficacy. The advertisement for Dr Nelson's panacea is typical of the hyperbolic claims of such advertisements, assuring its potential customers that:

> Dr. R. Nelson, being well known to have made the Cure of Seminal and Genital Imbecilities his chief Study and Practice for above 30 Years, does recommend his most Noble, Cleansing, and Strengthening ELIXIR, which thousands of People (many of them of High Rank) have happily experienced, and is by Numbers of Physicians and Surgeons approv'd as the only BALSAMICK, HEALING, and RESTORING Medicine, to be depended upon in the World. (*Universal Spectator* "Dr Nelson" n. pag.)

Similarly, the advertisement for the transcendent balsamick restrictive electuary claims:

When a medicine will infallibly accomplish such a safe, speedy, and perfect Cure of such difficult Maladies, as Gleets and Seminal Weakness are, as this Great Remedy truly and directly will, even after all other Means and Medicines have been tried in vain; too much cannot be said in it, and this, all who ever took it for any of the above mention'd Purposes, have readily declared. The price is but 6s. a Pot [and ...] is to be had only at Mr. Radford's Toy-shop at the Rose and Crown against St. Clement's Church-yard in the Strand, ready sealed up, with a Book of Instructions, which whomever carefully read, will be a complete Master of his own Case. (*Universal Spectator* "Concerning Gleets" n. pag.)

The rhetoric of these advertisements has much to tell us about both the popular concerns regarding "seminal weakness" and the status of medical practice at this time. Firstly, there is the assumption that prospective clients are likely to have already tried a range of other such tonics; secondly, that these treatments will be administered by the patient him or herself; and finally, that the commendations of those identifying themselves as licensed physicians will lend scientific credence to a tonic sold in a toyshop. These advertisements thus reflect a rhetorical invocation of medical knowledge as a guarantee of the effectiveness of the tonics they promote. At the same time, they also provide evidence of the existence of a whole industry catering to a clientele suffering (or believing themselves to be suffering) from sexual diseases and dysfunction: patients self-administer medications purchased in non-specialist commercial outlets.

The space mapped out at this intersection of medical discourses and commercial cultures proved an intensely productively one, at which new ideas (and profitable anxieties) about masculinity emerged and were exploited. We see this most clearly in the enormous impact of a text advertised alongside these advertisements in *The Universal Spectator and Weekly Journal: Onania; or, The Heinous Sin of Self-Pollution, And All its Frightful Consequences, in both Sexes* (then in its fifteenth edition). Like the handbooks published 150 years later by Kahn and Jordan, *Onania* served primarily as a "pamphlet and advertisement for nostrums" (Rosario 18), and anticipating these, it "swelled with new 'testimonial' letters from customers rescued from the deadly clutches of onanism" with every new edition (18). As Thomas Laqueur argues in *Solitary Sex: A Cultural History of Masturbation*, *Onania* represents a "shameless effort to invent a new disease" in order to "offer its cure at a steep price" (16). In this, it was wildly successful, and masturbation would continue to feature in nineteenth-century writing on "lost manhood" as one of the principle causes of both "seminal weakness" and the later condition of spermatorrhoea. As Thomas Laqueur has shown, while proscriptions on self-pleasure can be traced back to antiquity, masturbation itself is a specifically eighteenth-century invention, one tied to the emergence of modern understandings of subjectivity. Laqueur argues that the modern concept of masturbation reconceptualises earlier concepts of self-pleasure, understood as a sin of the flesh, as a corruption of the will and mind. Eighteenth-century anti-masturbation literature is less concerned to regulate bodily fluids or acts than it is about the subject's imaginative capacities, which have the potential to corrupt his/her rationality: "[t]hat which sullied, that which transgressed, was no longer semen in all its sticky specificity but rather the imagination revelling in desires of its own, unnatural creation" (216). The wilful turning of the conscious mind to voluptuous ends was seen as both a perversion of and, more worryingly, a danger inherent to, modern forms of a subjectivity that was understood to be rational and autonomous. In using the subject's imaginative capacities to base ends, Laqueur argues, solitary sex represented "the perversion of one of the mind's most protean and admirable faculties" (221).

In the nineteenth century, however, masturbation was decried not so much as a corruption of the will but as a failure of self-discipline. In a context in which, as we have seen, subjects were being taught by a vast interconnected network of popular and medical discourses to see their health as largely under their own control, as the direct result of the care and attention they dedicated to it, masturbation is a failure of the will that sullies the body and corrodes its health: one of the most common nineteenth-century terms for masturbation is "self-pollution." As both popular and medical texts in the nineteenth century emphasised, drawing on the long tradition of medical writing glossed above, the failure to exert self-control over the body's appetite for pleasure would debilitate its whole system. While this is continuous, again, with earlier ideas about the effect of seminal loss on the male constitution, it is also reimagined in specifically nineteenth-century terms as a loss of bodily self-discipline, or continence. In publications intended for a popular audience, like Kahn's *Treatise on the Philosophy of Marriage*, "the frightful and odious vice of self-pollution" (Kahn 75), was routinely identified as one of the most common causes of "lost manhood." The individual "afflicted" with a tendency to masturbate, writes Kahn:

> presents a melancholy and dejected appearance. He is restless, ever and again desiring a change, but disinclined to physical exercise. He seeks solitude, and by allowing his thoughts to dwell on the fact of his disability, frequently becomes hypochondriachal. In business he loses self-confidence, and is constantly in dread of some unforeseen event about to happen to him. His temper likewise becomes irritable, and he

is the subject of most sudden exacerbations of anger and passion.[...]
The healthy colour of the skin disappears, the eyes lose their brightness,
and are surrounded by a dark halo, while the state of the digestive organs
becomes exceedingly distressing. (81)

Commercial texts like Kahn's and specialist studies like Hunter's were
mutually reinforcing, the latter reproducing popular ways of seeing
and understanding the male body and sexuality, while providing a
legitimising framework for the former.

Despite their increasingly disreputable and marginal position
during the late 1800s, the influence of popular anatomical museums
on popular cultural assumptions about male sexualities and bodies,
and the spectacular culture to which they contributed, can be clearly
seen. Extending the discursive construction of health as a personal
responsibility, one productive of a respectable bourgeois subject, into
a commercial language, the proprietors of these museums helped
to medicalise male sexuality by reinforcing, and widely disseminating, worrying accounts about its dangers. Promiscuous or hedonistic
behaviours, men were told, would erode their virility and general
constitution: through excessive seminal losses they would lose their
masculinity altogether, becoming maudlin and melancholy, prone to
dizziness and blushing, lacking in confidence and energy. As Kahn's
comments above show, "lost manhood" was described as a spectacular
disorder; one of its distinguishing features was, precisely, that it caused
men to make spectacles of themselves, transforming their private vices
into a public shame. In this way, their very bodies provided evidence
of their cultural and corporeal transgressions, their failure to cultivate
the discipline and moderation required of the Victorian patriarch.
Public anatomical museums "for men only" played a significant role
in popularising this view by spectacularising what happened to men
who deviated from these norms and practices, while warning that
the conditions of modern life afforded men constant temptations to
destroy their own health and rectitude.

At the same time, they also provided anatomical evidence for the
need to constantly survey one's own body and practices, and to work
on and improve its general health. We see this in the emergence of
the new medical condition that came to eclipse what was previously
called "seminal weakness" (in professional medical writing) or "lost
manhood" (in museum handbooks and advertising): spermatorrhoea.
In a context in which the changing economic and domestic conditions
of urban life were seen to be softening men to the point of rendering them incontinent, entirely liquid, spermatorrhoea offered a new
set of diagnostic criteria and an anatomically based cure for men

whose masculinity had been damaged or become diseased. What the spermatorrhoea epidemic of the late nineteenth century demonstrates is how continuous the exhibitory techniques used in these increasingly lowbrow museums were with the medical construction of the spermatorrhoeaic subject as spectacularly decaying and degenerate, which is found in the professional literature of the same period.

Spermatorrhoea and the Spectacularised Male Body

The 1863 edition of "Dr" Kahn's *Treatise on the Philosophy of Marriage*, included as a supplement to the *Handbook of Dr Kahn's Museum*, followed a common practice in much nineteenth-century medical writing — both professional and popular — in including letters from his patients as part of his own text. These provided first-person accounts of the patient's illness, as well as testimony to the efficacy of his treatment. One letter writer outlines his problem to Kahn in the following terms:

> I am twenty-seven years of age, of a delicate, nervous temperament; I am single, and likely to remain so, unless you can assist me; for there is no disguising the fact, I am *impotent* through the effects of self-pollution, which I practised from eleven years of age until twenty-two, when I became acquainted with its mischief and left it off *for ever*. I then obtained medical advice, which gave me only temporary relief.[...] I am much afraid I am suffering from *Spermatorrhoea*.[...] I have a slight cough always on me, with shortness of breathing, and I am very thin. I often turn very giddy when rising or stooping hurriedly. Reading the slightest thing of a sentimental character brings tears to my eyes, which I cannot help, although I feel them to be maudlin.[...] I have no confidence in myself. I blush and look guilty at the slightest thing said to me, whether right or wrong; blushing and becoming pallid by turns. (180; original emphasis)

Letters like this were sent in their thousands to popular medical figures like Kahn during the second half of the nineteenth century and the first decades of the twentieth, written by men worried that they were suffering from what the "museums of anatomy" described above referred to as "lost manhood," but which was in the mid-nineteenth century identified as a medical condition, "spermatorrhoea."[9] As the letter cited above indicates, spermatorrhoea was identified not simply as a genital disorder, but as manifesting in a broad spectrum of secondary symptoms: crying, blushing, sensations of guilt or sadness, lack of confidence or anxiety, and so on.[10] As the letter cited above indicates, the problems associated with spermatorrhoea — and, indeed, the onset of the condition itself—were largely held to be a consequence of men's own behaviour. Men who indulged in intemperate practices such as masturbation were suffering from lapses of physical and/or mental self-control that would certainly destroy their virility but would also jeopardise their entire constitution. Such men, suffering from "lost manhood," would find themselves not only impotent but also incapable of assuming their proper patriarchal role as men.

In this respect, spermatorrhoea might be understood as a male equivalent of hysteria, which is also associated with a wayward sexuality and non-normatively gendered behaviour. And, like hysteria, spermatorrhoea was, clearly, a contagious idea, addressing pre-existing anxieties about masculinity to which readers were predisposed to be susceptible. It quickly became the subject of an epidemic of diagnoses. However, in contrast to hysteria, which has been the subject of much study by medical historians and feminist scholars, spermatorrhoea occupies a comparatively very obscure position within the history of medicine and masculinity studies. Yet for Victorian physicians like Albert Hayes, director of the Boston Peabody Medical Institute, spermatorrhoea was among "the most dire, excruciating and deadly maladies to which the human frame is subject" (1). Moreover, the fact that members of the general public were already familiar with the term and the diagnostic criteria for spermatorrhoea in the 1860s, soon after the condition had first been identified, and that they were seeking advice from the proprietor of an anatomical museum for its treatment, is a telling indication of the role these museums played in late nineteenth-century culture. Despite their declining cultural status and increasingly geographical marginality, then, late nineteenth-century anatomical museums remained a part of the vast network of health institutions and discourses, reinforcing the widespread understanding of health as a consequence of private behaviours and a measure of personal self-discipline. Museums like Kahn's served an important if increasingly disreputable role in this cultural context by vividly spectacularising the unhappy fate that awaited the undisciplined body. Such men, suffering from the loss of their manhood, would find themselves not only impotent but also incapable of assuming their proper patriarchal role as men.

The term spermatorrhoea, or *spermatorrhée*, was coined in 1836 in the first volume of the French physician Claude François Lallemand's *Des pertes séminales involontaires*, and was used to refer to excessive discharges of semen. Lallemand introduced this newly identified condition by explaining: "Neologism is only excusable when it has as its goal the desire to anticipate and prevent error [...] And so, to avoid awkward paraphrases I will call all excessive seminal evacuations, of whatever kind they might be, *spermatorrhée*" (7).[11] Like onanism in the eighteenth century, "spermatorrhoea" represented something both new and not new in the nineteenth century: in identifying this disorder, Lallemand's text did not invent the cultural anxiety surrounding male bodily fluids and the impact of their loss on the health and vitality of the male body, but he did reimagine the relationship between fluidity and the male body in a way that reformulated the cultural significance of seminal fluidity itself. As Peter Brown notes, for medical writers in Aretaeus's time (and for long afterwards), ejaculation was seen as dangerous because it so closely resembled the convulsions of epilepsy—"Did not the very mouth of the epileptic also froth with the same bubbling, whitened blood as did the penis?" (18). The ooziness and seepage associated with spermatorrhoea belongs to a new and distinct imaginary of male bodily fluids, however, specific to the nineteenth-century concerns about a softening, degenerate, over-civilised masculinity. During what she has termed "the spermatorrhoea panic," Ellen Bayuk Rosenman writes, "semen was pathologised as the symbol of everything that is alarming about the body; [...] it becomes a 'thin, imperfect fluid' that no longer spurts majestically but 'dribbles from the end of the penis.'" (373). That is, spermatorrhoea reimagines ejaculation not as a spasmodic convulsion, as it was described in almost all earlier writing, but as an incontinent, seeping leakage.

To the list of symptoms originally identified by Aretaeus as consequent to "excessive" ejaculation, Lallemand added those recognisable only to the trained eye of the anatomist: the post-mortem identification of testicular asymmetricality and other genital pathologies, and, more particularly, the existence of spermatozoa in the urine, detectable only through the new medical technology of microscopic analysis, which soon became the main diagnostic criteria of spermatorrhoea. Thus, just as eighteenth-century anatomists saw in the reproductive anatomy of the female body evidence of, and justification for, emerging ideas about femininity and maternity, so did spermatorrhoea provide a new diagnostic category in which nineteenth-century concerns about masculinity, virility and self-control could be read in the sexual anatomy of the male body. "Excessive" seminal losses were now said not only to deplete men's reserves of virility but also to produce actual genital pathologies. Indeed, Lallemand's first case studies, and his identification of spermatorrhoea, derived from his post-mortem analysis of testicular asymmetries and pathologies.

What Lallemand's research thus demonstrates is the extent to which the anatomist's gaze is shaped by pre-existing cultural assumptions about the body, seeing in the anatomies of men evidence of their masculinity. Lallemand may, in this respect, seem an eccentric or extreme case, even in nineteenth-century medicine; however, as Michael Mason notes, although "Lallemand's work reads nowadays like the production of a madman" (298), he was nevertheless "a competent medical scientist (he contributed elsewhere to the accurate understanding of spermatogenesis), and in an age when almost all pathology was a matter of syndromes [...] the grotesque constellations of sexual history and symptoms which are Lallemand's 'cases' were not obviously unscientific" (298). One proof of this is how quickly Lallemand's work was taken up by the mainstream medical profession. In the years immediately following the publication of *Des pertes séminales involontaires*, dozens of texts on spermatorrhoea—by registered physicians and popular writers alike—appeared, and these continued to be published into the twentieth century (although the existence of the condition had long been medically debunked by then). These included Dr Pickford's *On True and False Spermatorrhoea* (1852), T. H. Yeoman's *Debility and Irritability Induced by Spermatorrhoea; the Symptoms, Effects, and Rational Treatment* (1854), John Milton's *On Spermatorrhoea: Its Pathology, Results, and Complications* (1856), Albert Hayes' *The Science of Life; or, Self-Preservation. A Medical Treatise on Nervous and Physical Disability, Spermatorrhoea, Impotence and Sterility* (1868), and M. K. Hargreaves' *Venereal and Generative Diseases including Disorders of Generation, Spermatorrhoea, Prostatorrhoea, Impotence and Sterility in Both Sexes* (1887). Proprietors of anatomical museums and sellers of patent medicines such as Kahn and Jordan were also very quick to adopt this term, writing at length about spermatorroea in their museum handbooks and publicising the dangers of this condition in mail-order circulars and publicly displayed handbills.[12] Additionally, and as the letter to Kahn cited above demonstrates, members of the general public were quick to identify themselves as suffering from this disorder.

My own concern is thus less to appraise whether Lallemand's work was legitimately scientific than to examine the interrelationship between anatomy and spectacularity, between medical knowledge about the male body and popular assumptions about masculinity. The spermatorrhoea epidemic, and the literature about it, whether written by respected medical practitioners like Lallemand or publicly denounced charlatans like Kahn, addressed anxieties about masculinity to which nineteenth-century men were predisposed to be susceptible, and it is this that gave the disorder such resonance in the popular imagination of the time. The extent to which Lallemand's work catalysed already existing concerns about masculinity and the sexual male body can be seen in the extraordinarily rapid uptake of his work. As a category, spermatorrhoea proved capacious, absorbing everything previously associated with "seminal weakness" and further multiplying its symptoms. It is Lallemand's neologism of *spermatorrhée* that initiates what is certainly the most intensive period of writing on and treatment of "seminal losses" on record.[13] Neither Hunter's text, nor that of Ernest Wichmann (whose 1782 *De pollutione diurna frequentiori sed rarius observata tabescentiae caussa* Lallemand cites as the only previous study in this area), sparked anything like the discursive explosion that followed the publication of Lallemand's book. In this respect, although Lallemand's text clearly and identifiably emerged out of the longer history of writing of seminal loss traced above, his neologism nonetheless did produce something genuinely new: not just more writing on this established subject, nor even the new treatment practices that were so widely developed and practised in its wake, but a peculiarly nineteenth-century reconceptualisation of the relationship between masculinity, medicine, and male body fluids closely and identifiably informed by the culture of anatomical museums. Spermatorrhoea is hence important for the new ideas about masculinity it produced and the new ways of seeing the male body it popularised.

Spermatorrhoea is, moreover, a rare case in which ideas about fluidity are associated with the male body rather than the female. As Elizabeth Grosz argues, "the fluid, the viscous, the half-formed" is almost always aligned with the female body (195), and discussions of male bodily fluids remain extremely rare. This, she argues, reflects "men's attempt to distance themselves from the very kind of corporeality—uncontrollable, excessive, expansive, disruptive, irrational—they have attributed to women" (200). It is just these characteristics that we see associated with spermatorrhoea: and this is the reason spermatorrhoea is identified as a pathological condition and associated with men's feminisation. We see this in the descriptions of the key symptom of spermatorrhoea as an incontinent leakiness. As Ellen Bayuk Rosenman argues:

> In its obsession with the breakdown of continence, spermatorrhea literature provides an encyclopedic rendering of nonnormative masculinity. These men suffer from paralysis, tremors, lassitude, and insomnia; they cannot concentrate, work, or get out of bed; they are nervous, weepy, distracted, afraid. Spermatorrhea offered a dumping ground for all the traits that bourgeois masculinity had to abjure, traits associated with lack of control over the body and the emotions and with a loss of confidence and power. (375)

Causes of the disorder were seen to vary widely, but were generally attributed to an overly domesticated and unmanly lifestyle: feather beds, soft trousers, excess reading, sentimental literature, and sedentary pursuits were all cited as possible causes. However, most physicians agreed with Robert Bartholow that "the vice of masturbation is undoubtedly the chief cause" (5). In *Spermatorrhoea: Its Causes, Symptoms, Results and Treatment*, Bartholow wrote:

> although the practice of this vice is not confined to boys of the nervous type, yet it finds in them victims the most willing, and the least able to resist the continually increasing demands of the habit. Boys of vigorous constitution, in whom the digestive and muscular systems are well developed, are less under the control of these erotic impulses, and are more able to resist the inroads of the habit when formed, because in them the exercises of youth and the satisfaction of appetite occupy their minds. (6)

As for Bernarr Macfadden, for Bartholow boys with generally good practices of hygiene and bodily self-discipline will be the best prepared to exercise self-control, thereby assuring their adult health and preparing themselves for the responsibilities of bourgeois Victorian life.

Museums of anatomy and medical literature of the late nineteenth century again played an important role in reinforcing this message by offering up for public scrutiny empirical evidence of the pathologising effects of masturbation on the male sexual anatomy. Kahn, for instance, warns:

> From the continual excitement and constant action of the parts in the formation and emission of large quantities of seminal fluid, the veins become enormously distended, and apparently more numerous, and their coats thickened; the scrotum generally becomes elongated on the affected side, more frequently the left, but sometimes both; the folds disappear, and the whole organ hangs down in a pendulous state; sometimes the testes waste entirely away, and, as a matter of course, impotence, in many cases incurable, is the result. (77)

This passage includes reference to exhibition item numbers that provided visible and anatomical proof of Kahn's claims, the incontrovertible evidence of masturbation's pathological effects.

In "Self-Diagnosis; or, How shall we ascertain under what affection we are suffering?" Kahn provided a list of the symptoms by which one might identify spermatorrhoea that ran to over three pages, categorised as local (genital), general (bodily), or mental (*Handbook of Dr Kahn's Museum* 120–23). In this way, the symptoms of spermatorrhoea, like those for "seminal weakness" and gonorrhoea before it, extended from the slackening of the sexual anatomy into a general laxity of behaviour and deportment. In addition to "semen accompanying urine or defecation," localised symptoms included both "erections and emissions upon slightest excitement" and "a decrease of sexual desire or enjoyment," both nocturnal and diurnal emissions, "priapism" and "diminution of penile size," "emissions without erection" and "difficulty maintaining erection." Similarly, generalised bodily symptoms encompassed both signs of an overheated system, such as "quickened pulse," "flushed face," "irregular heart beat" and "diarrhoea," but also signs of sluggishness, such as "cadaverous appearance of skin," "hollow or sunken eyes," "extreme sensitivity to cold," "lassitude," "fatigue on slight exertion" and "constipation." In addition to these were the mental effects of spermatorrhoea: "restlessness, sighing, want of energy, uncertainty of tone of voice, nervous asthma, vertigo, want of purpose, aversion to society, blushing, want of confidence, avoidance of conversation, desire for solitude, listlessness and inability to fix the attention, cowardice, depression of spirits, giddiness, loss of memory, excitability of temper, moroseness, want of fixity of attention, lachrymosity, strange or lascivious dreams and hyprochondriasis." Kahn's list finishes in ominous capitals: "CLIMAX: INSANITY" (120–23).

As Kahn's rather histrionic tone made clear, the symptoms of spermatorrhoea, and the consequences of failing to discipline one's sexual practices, were seen to manifest themselves in spectacular fashion. While this is perhaps unsurprising in popular or quack medical texts, it should be recognised that professional medical literature of the same period viewed the sexual male body through a very similar lens. Spermatorrhoea renders public and shameful the subject's private loss of self-control, his inability to live up to the expectations of dominant nineteenth-century masculinity. Although Kahn—along with subsequent public anatomical museum proprietors like Jordan and the Reinhardts—had an obvious commercial motivation for his sensationalised rhetoric, it is important to recognise that this was entirely consistent with the tone of contemporaneous professional medical literature. John Skelton's *A Treatise on the Venereal Disease and Spermatorrhoea* describes spermatorrhoea in terms less melodramatic than Kahn's but no less severe:

> The symptoms of spermatorrhoea are very various; and as the disease advances, the mental condition of patients generally undergoes a marked change. They become fretful and peevish; their memory fails; they lose their courage, and indignities, which they would formerly have resented, they now endure with patience. Occasionally it assumes a much more serious aspect, and they become confirmed hypochondriacs; are unfit for either business or serious reflection, and are disagreeable to themselves and the whole world. (108)[14]

As such claims demonstrate, medical accounts of the spermatorrhoeaic body are not free of elements of spectacularisation in the way they see that body, and in this way they are intricately and closely interconnected with the exhibitory cultures of their time: in holding a particular kind of body up to public scrutiny as the exemplar of its type, they make a medical exhibit out of he who makes an exhibition of himself.

Deviating from the norms of late nineteenth-century masculinity, lacking self-governance, sufferers of spermatorrhoea were subjected to treatments that were explicitly and forcefully disciplinary. These usually followed one of two main approaches. The first was to focus on improving the general health and vigour of the body: "Few means of controlling spermatorrhoea could be devised so simple and natural as exercise, especially gymnastics," wrote John Milton (*Practical Remarks* 6). The patient was encouraged to participate in his own treatment and might "do half the surgeon's work if he will rise at five or six o'clock, sponge with cold salt water, use the dumb bells for half an hour, and follow this up with a brisk walk. It will not be long before the eye grows brighter, and the skin clearer; before he sleeps sounder and again feels comfort in existence" (6). Skelton concurred:

> The patient should be put under a course of tonics, he should habituate himself to cold bathing, [...] where sea bathing can be had it is prefer-able.[...] The mind should be employed on some useful occupation, so that it may not brood over the disease, carefully avoiding, however, anything requiring very close application, for severe study is often a cause of seminal discharges. (116)

Unsurprisingly for a disease characterised by a slackening of body and will, the simplest remedy was seen to be to subject the body to a series of disciplinary practices which, like those recommended by popular health advocates such as Bernarr Macfadden, were designed to toughen and harden the body. In a context in which the public was warned that they were personally responsible for the condition and cultivation of their own health as a means of self-improvement but also as a social obligation, disciplining the unruly body seemed the most logical cure for a self-induced condition like spermatorrhoea.

If self-discipline failed, however, medical intervention was deemed necessary, and the severity of these interventions demonstrates how dangerous spermatorrhoea was seen to be. Robert Bartholow's pre-ferred remedies included the application of a "*porte caustique* to the prostatic portion of the urethra and the use of injections of the nitrate of silver, sulphate of copper, acetate of lead, etc.; certain mechanical expedients; [...] cold hip-bathing, and injections of cold water into the rectum" (78–79). Additional treatments included acupuncture of prostate and testes, blistering of the penis, and forced dilation of the anus. "I have had excellent results from stretching the sphincter ani,"

Bartholow wrote. "The method as pursued by me consists in the introduction of a bi-valve rectal speculum, and then working the screw until the blades are sufficiently separated. The operation causes considerable pain, and may rupture the sphincter if incautiously carried too far [...] but it has seemed the most useful in the cases of simple spermatorroea" (93–95). The brutality of these treatments, and the seemingly self-defeating policy of attempting to cure the "irritation" and "over-excitement" of the genitals by inflaming and irritating them still further, attests to a strong determination to discipline the male body, in order to prevent its dissolution into a pathological ooziness. Medical treatment is here, clearly and explicitly, operating as a disciplinary technology: "Spermatorrhea cures seem shockingly invasive, even punitive," Rosenman notes. "It is as if, in violating the fantasy of continence, the body has forfeited any claim to intactness; its fate is to give up its boundaries altogether to the doctor" (376). In failing to live up to a dominant cultural ideal of masculinity and the male body—as rigidly self-contained and hermetically sealed—the liquid spermatorrhoeaic body must be disciplined and punished through the very orifices that are the source of its pathological excesses.

The medical treatments for spermatorrhoea in this way make explicit the threat contained in the curatorial style of the popular museum of anatomy in this period—the body that fails to discipline itself and to manage its own bodily practices will not only become diseased but will also be punished, both socially and physically. At the same time, however, the rapid spread of spermatorrhoea in the second half of the nineteenth century cannot be simply attributed to its imposition as a diagnostic category on reluctant or resistant men; rather, as we saw in the letter cited at the start of this section, men were very quick to self-identify as spermatorrhoeaics and to request (sometimes very severe) medical help in regaining control over bodies that seemed wayward or unruly.[15] Benjamin Phillips writes in an early article on spermatorrhoea for the *London Medical Gazette*: "Since the publication of the first part of this paper, I have been painfully impressed with the conviction that the evil is more widely spread than I had before conceived.[...] Almost every morning I have had several applicants for relief, but with two or three exceptions they have been either medical men or medical students" (315). In other words, the men initially identifying as spermatorrhoeiacs were those reading the emergent literature on spermatorrhoea. As discussion of spermatorrhoea radiated outwards from professional medical journals to the popular press, the number of men who became convinced they were suffering from the disorder reached epidemic proportions.

Ian Hacking has described the relationship between patients and medical discourses as one characterised by a "looping effect" (in "Making Up People," and "The Looping Effect of Human Kinds"). Examining how classifications "affect the people classified, and how the effects on the people in turn change the classifications," Hacking argues that although we think of "kinds of people as definite classes defined by definite properties," they are in fact "moving targets because our investigations interact with them, and change them. And since they are changed, they are not quite the same kind of people as before. The target has moved. I call this the 'looping effect'" ("Making Up People" 224). In a similar way, although the letters from patients to their doctors might seem to ventriloquise the discursive construction of male bodies in medical literature at this time—indicating that the incitement to internalise the information on display in popular museums of anatomy and medical literature was highly successful—we must also remember that these letters were published in the context of texts that had been edited by their recipients. As Hacking argues, in the complex institutional and discursive network in which these ideas circulated, looping effects do not simply reproduce, but transform the cultures in and through which they take place. Perhaps the most significant example of this is the way the career of spermatorrhoea as a medical disorder produced actual institutional change in the practice of medicine and its professional organisation.

Although spermatorrhoea itself is now a rather obscure footnote in the history of medicine, it is as a direct result of this epidemic—and especially its commercialisation in and by the quack medical institutes associated with "museums of anatomy"—that professional medical practice was extended to include the treatment of sexual diseases and genito-urinary specialisations for the first time. As Angus McLaren notes in his cultural history of impotence, urology was for a long time "tainted by its association with venereal disease and impotence," and doctors "who discussed such issues were acutely aware of their apparent unseemliness. Moreover, the terrain was already occupied by quacks" (127). In consequence, as Haller and Haller remark: "Many hospitals in New York and elsewhere had rules prohibiting the treatment of gonorrhea or syphilis" (263). The cultural bias against those who had contracted sexually transmitted diseases and the refusal to provide for their treatment, was, once again, a product of the belief that subjects had an obligation to take responsibility for their own bodily health and management: "Venereal diseases were contracted, usually, through voluntary decisions," David Pivar explains. "It was popularly thought the diseases were a punishment for transgression.

If not treated, the fear of disease might be a deterrent" (106). It is in the spermatorrhoea literature that we see a concentrated effort to challenge this and to make the treatment of sexual disorders a part of the practice of mainstream medicine. The mid-nineteenth century campaign against quacks like Kahn also represented a campaign for the development of genito-urinary specialisations. In *Practical Remarks on the Treatment of Spermatorrhoea and Some Forms of Impotence*, John Milton writes that:

> it has always appeared strange to me that this affection should remain abandoned by the profession to a few solitary specialists, and for the benefit of the vile harpies who prey on this class of victims.[...] This neglect, and the twofold indisposition of the patient either to trust his ordinary medical attendant with the secret of his disease, or to permit him to exercise that operative interference which the specialist will perhaps insist on; the extent, the manner in which the question has been studiously burked, and the absence of any *well-known* source to which he can turn for information, have had the natural effect of driving him to those who will make it their business to let him know, that so long as he has money there is one city of refuge to which he can always fly (2; original emphasis).

Dr Pickford concurred: "It is [...] this inexcusable neglect in medical men, which drive[s] the [sufferer] into the hands of nostrum-vendors and infamous quacks" (6). Bartholow, too, claimed that "it is a reproach to our profession that this subject has been permitted, in a measure by our own indifference, to pass into the hands of unscrupulous pretenders, whose suggestive publications are among the crying evils of our time" (iii). He noted that, because "the subject is disagreeable, and to a certain extent disreputable, competent physicians are loath to be concerned with it.[...] For this reason, and to obviate the sad consequences which result from spermatorrhea, it is our duty to exert our best efforts on behalf of those afflicted with this malady" (iv). An editorial in an 1857 edition of *The Lancet* further urged: "Let honourable and scientific men take possession of the field now occupied by these vagabonds" (251).

This self-representation of a medical profession reluctantly turning to the neglected and distasteful disease of spermatorrhoea in order to save suffering men from the dangerous ministrations of quacks is primarily a rhetorical strategy, it should be emphasised: mainstream doctors and quacks offered similar, and sometimes identical, treatments. However, it is a rhetoric that was mobilised in the interest of affecting concrete institutional change, strengthening the professionalisation of this area of medical practice by prompting legal action to formally exclude and delegitimise the practice of quack doctors and restructuring general medical practice to include the treatment of sexual diseases, disorders and dysfunctions. Treatment of such conditions would no longer be the province of medical entrepreneurs, but would be taken over as the exclusive terrain of licensed professionals. Patients would no longer treat themselves with store-bought medical preparations, but would submit themselves to the treatment of physicians using new medical approaches and technologies. In this way, spermatorrhoea produced not just a greater volume of discourse on the old topic of seminal incontinence, not simply its translation into the context of specifically nineteenth-century cultural concerns, but also a profound institutional shift in the structure and practice of medicine.

By the early 1860s, a spate of texts on "true and false spermator-rhoea" began to emerge, in which "false spermatorrhoea" was identified as that diagnosed by quack doctors, and "true spermatorrhoea" was redefined as a much rarer condition only a licensed physician could detect. This signalled the beginning of a rapid decline in spermator-rhoea diagnoses, and within a few short years, this epidemic had died away as quickly as it flared up. Having transformed what had previously been known as "secret diseases" into something understood under the rubric of "sexual health," and having produced a series of corollary structural changes in the profession and practice of medicine, spermatorrhoea appears to have served its cultural purpose. As such, this condition represents both the culmination of the long history of medico-moral writing on seminal incontinence and, simultaneously, its end. The effects of this history, however, would endure for a considerable period afterwards. The public campaigns for sexual health that began during the First World War in the USA—initiated by the American Social Hygiene Association, a government agency that had developed from the earlier purity movements—called "Fit to Fight" (aimed at mobilised soldiers) and "Keeping Fit" (for civilians) retained this emphasis on the need to keep one's body disciplined and strong, and the importance of resisting appetites that would destroy the body, ruining its own health, and that of the family for which it was responsible and of the nation as a whole. Like Bernarr Macfadden—whose journal *Physical Culture* reviewed the film of *Fit to Fight* and called for its release to a civilian audience—early twentieth-century proponents of fitness saw this not only as a physical but also as a moral attribute, thereby reinforcing the fundamental tenet of contemporary health discourses, that work on the body is a form of work on the self.[16] The unfit body, like the body suffering from spermatorrhoea, makes visible on its surface the damage done to its internal anatomy through a lack of proper care and cultivation. In a period in which bodies were being measured and assessed by a range of professions—including phrenology, criminal anthropology, ethnography and, of course, medicine—the body's external form was widely read as a text, the legible account of its hidden interiority. Nineteenth-century audiences, in both professional and popular contexts, were encouraged to see the body as a site of spectacularisation of the self and its behaviours, made possible by the anatomical vision of the body that the spermatorrhoea epidemic had further lodged in the public imagination.

In this way, the "lost manhood" industry of the turn of the nineteenth century represents a continuation of longer traditions of writing about seminal fluidity and their transformation within a new context, in which medical literature, mass-marketing, new visual technologies and exhibitory cultures come together in a new and profoundly influential way. As such, even after spermatorrhoea itself had disappeared as a medical diagnosis, the ways of seeing the male body and the knowledge about that body it popularised continued to exert an important influence on public discourse about male sexuality and masculinity well into the twentieth century.

Notes

1 In its appeal for Jordan's museum to be closed, *The Age* cited the recent closure of Kahn's museum in London—although they appear to have been unaware of Jordan's involvement in this latter site. (For coverage of this case, see Robertson 164–80).

2 As Armstrong and Armstrong note of the role of regulation in nineteenth-century American medicine: "Even the casual standards of early medical education were too restrictive for democratic health reformers, who proclaimed that in a free country, anyone should have the right to practice medicine. In the 1830s, a groundswell of sentiment called the popular health movement temporarily ended nearly all government regulation of health care, part of the wave of Jacksonian anti-elitism. By 1845, only three states still licensed medical doctors" (3).

3 The most profitable of these, Brinkley's Formula 1020, was sold at a mark-up that Pope Brock calculates to have been 9,200 per cent (243).

4 Brinkley himself had some medical qualifications, having begun, but never completed, medical studies at several private colleges in the USA.

5 The purity movements recommended vigilant self-governance of personal (and especially sexual) hygiene as a means of minimising the social and sexual ills of the period, focusing primarily on prostitution and venereal disease. In the early 1900s, the purity movements were gradually restructured and replaced by newly formed government organisations such as the American Social Hygiene Association. (See, for instance, David Pivar's *Purity and Hygiene: Women, Prostitution, and the "American Plan," 1900–1930*).

6 The Swiss physician Samuel Auguste Tissot (1728–1797) was not a quack doctor but a well-regarded neurologist, as well as a physician, who remains widely

recognised for his pioneering work on the study and treatment of migraine (in *Traité des nerfs et de leurs maladies* [*Treatise on the nerves and nervous disorders*]).

7   Ejaculation has often been a source of cultural anxiety, because it is an act which simultaneously fulfils the phallus's promise of fertility and deflates it into the flaccid penis, thereby revealing the instability and fluidity at the heart of a masculinity constructed as rigidly fixed and self-contained. Ejaculation is thus something of a constitutive paradox in phallic masculinities, providing both proof of its virility and an exposure of its disavowed fluidity.

8   In emphasising this aspect of gonorrhoea, Aretaeus's position is somewhat different to Galen's. Although Galen also described gonorrhoea as a pathological condition, he was more concerned with the detrimental effects of "accumulated seed," which could "petrify" within the body if not periodically and therapeutically released (see Rosario 16). However, Galen's view remained a much more marginal one than that of Aretaeus, which would set the tone for writing in this field for fifteen hundred years.

9   The guides to male sexual health included in Jordan's museum handbooks also included letters from his patients. These letters were very similar in style and address to those published by Kahn. A typical piece of correspondence reads: "I am now twenty-four years of age, stand five feet nine inches in height, and weigh ten stone. My general health is good; but my appetite fails me very often; [...] sometimes I feel quite melancholy; at other times, I have a superabundance of animal spirits. I am not much subject to pain, but sometimes after I have committed the act, I have felt a slight pain in the left testicle, and at other times in the passage of the yard; at other times, I have felt a dull pain in the left side for two to three hours at once. I am subject to frequent and nightly emissions; in voiding urine, I have seen seminal fluid run away from me thin and unelaborated, especially when straining out the last few drops, and the end of the yard is almost constantly wet with fluid that escape from me.[...] I feel often very dull and heavy about the head as though from determination of blood; and my superintendent often taxes me with being careless, when at the same time, it is through an impaired memory. I sometimes forget myself altogether, as being unconscious of persons being present." (Jordan and Davieson 41–43).

10   Although such symptoms suggest spermatorrhoea was closely aligned with concerns about neurasthenia in the late nineteenth century, this latter was a disorder of the nerves and brain, whereas spermatorrhoea was associated directly with virility and the genitals.

11   Caude François Lallemand (1790–1853) trained at the military hospital in Metz and under Guillaume Dupuytren (whose collection of anatomical specimens later became part of the Musée d'Anatomie Delmas-Orfila-Rouvière) in Paris. He completed a doctorate in medicine in 1819, and was appointed professor of Clinical Surgery at the Medical School of Montpellier in the same year. In 1845, he was elected a member of the Paris Academy of Sciences.

12   The uptake of the idea of spermatorrhoea was particularly strong in the United Kingdom and, later, the United States. Both these countries witnessed the publication of vastly more texts on spermatorrhoea than in France, and in both the UK and USA the effect of this writing on medical treatments for male sexual disorders was much more immediate and pronounced. Texts dedicated to the study and treatment of spermatorrhoea in France do not appear until half a century after the publication of Lallemand's work, and appear to be influenced by the subsequent importance this work has acquired within British medical practice.

13 Beyond his neologism of the word "spermatorrhoea," Lallemand is widely acknowledged to have single-handedly initiated the intense period of medical investigation and practice in this area in the mid-nineteenth century, both in contemporaneous medical literature and in subsequent historical accounts. John Milton, surgeon to St John's Hospital for Diseases of the Skin in London, and himself author of some of the first texts on spermatorrhoea published in English, typifies the reception and importance accorded Lallemand when he argues that "to M. Lallemand belongs the merit of having forced upon the profession a recognition of the importance of the disorder" (*On Spermatorrhea* 5). Milton's first full-length text on spermatorrhoea, *Practical Remarks on the Treatment of Spermatorrhoea and Some Forms of Impotence*, appeared in 1854.

14 The prevalence of Skelton's account of the symptomology of spermatorrhoea is reflected in the fact that it is repeated, almost word for word, in Richard Dawson's *An Essay on Spermatorrhoea*. For Dawson, one may recognise a sufferer by the way his "general deportment undergoes a remarkable alteration. His temper, for instance, is extremely irritable, and he is fretful, peevish, discontented, and his appearance shows a marked degree of melancholy. Such patients are far from being courageous, or excited to anger or resentment, even by those incidents which, other under circumstances, would arouse their indignation. On the contrary, they are timid, fearful and apprehensive, and endure injuries which they have neither the spirit nor courage to resent" (4).

15 The extent to which men were willing to subject themselves to long and often brutal courses of treatment is revealed by the surgeon Thomas Curling's statement that patients "troubled with seminal emissions which no effort of the will can prevent their provoking, or which persist in spite of medical treatment, have in some instances been solicitous for the removal of their testicles, to get rid of the disgusting complaint; and individuals have even been known to perform the operation of castration on themselves" (403–404).

16 The American Social Hygiene Association, with its emphasis on keeping oneself "clean" and "pure," was also closely related to the eugenic movement; one of the ASHA's key organisational figures was Francis Galton, who coined the term "eugenics."

CHAPTER 3

From the Freak to the Disabled Person: Anatomical Difference as Public Spectacle and Private Condition

When P. T. Barnum staged his first exhibition in 1835, advertising an African-American woman, Joice Heth, as the 161-year-old former nursemaid of George Washington, he established what would develop very quickly into a new and hugely popular form of public spectacle, the freak show, and in so doing he also reinvigorated public interest in exhibitions of human anatomy as a form of popular entertainment.[1] While the freak show's emphasis on living exhibits might seem categorically distinct from the anatomical exhibitions examined in the previous chapters, which focused on human remains or artificial models, freak shows were (and remain) a closely connected part of the wider exhibitory culture examined in this book. Barnum named his first permanent site the American Museum and, as president of the New York World's Fair in 1853, played a pivotal role in establishing a cultural continuity between the culture of these official events and that of commercial exhibitions.[2] Coinciding with the period in which the culture of public spectacles and large-scale exhibitions was experiencing its highest levels of popular and institutional support, Barnum's exhibitions originally occupied a much more respectable cultural position than their later reputation would suggest. As James Cook recognises, although exhibits such as the "What is it?"—a long-running freak role played by a succession of microcephalic (or "pinhead") performers—would later

> be confined to more carefully segregated lowbrow venues (the carnival midway, or the seaside amusement park), during the 1860s What Is It? was still considered solid family entertainment, an exhibition worthy of visits by the upper crust as well as respectable workers and women and children, as well as men, serious naturalists as well as fun-seekers. (140)

During the same period in the UK, commercial exhibition venues such as the Egyptian Hall, the Cosmorama Rooms, Regent Gallery and Saville House were frequented by medical professionals, government ministers and members of the aristocracy, along with the general public. Reporting on the 1824 London exhibition of Caroline Crachami (the "Sicilian Dwarf"), The *Morning Chronicle* noted, similarly, that "the morning calls of the Royal Family, the Nobility, the Foreign Ambassadors, and the highest members of the Faculty, and others of rank and fortune, have frequently exceeded two hundred" (3). Popular performers such as General Tom Thumb (a dwarf), Millie-Christine (African conjoined twins) and Julia Pastrana ("The Missing Link") toured internationally and elicited a great deal of favourable press coverage [Fig. 12].[3]

As this wider culture indicates, Barnum was hardly the first to exhibit people with rare congenital conditions—such as conjoinment, hirsuitism or elephantiasis—as a form of popular entertainment. Exhibitions of human "monsters" or "curiosities" have a much longer history than that with which this book is concerned. However, the advent of the word "freak" as a new term for anatomically unusual forms of embodiment is representative of a new nineteenth-century tendency to explicitly encourage audiences to see those bodies through the lens of medical knowledge.[4] On the souvenir cards Barnum and other showmen had printed of their performers, portraits (made with the new visual technology of photography) were supplemented by medicalised accounts "by physicians who had examined the performer, declaring his or her malformations to be genuine" (Dennett 77).[5] While such text was clearly included for commercial and rhetorical purposes—providing evidence of the exhibits' authenticity and legitimate scientific value as much as an explanation of their particular medical condition—it is also true that medical science did take a keen professional interest in the anatomically unusual bodies displayed in commercial contexts. This is reflected in the fact that the medical study of teratology developed over the same period that the freak show flourished as a popular spectacle.[6]

The most significant indication of the close relationship between scientific and spectacular cultures within the space of the nineteenth-century freak show is the fact that Barnum's exhibitory practices extended to actual anatomical dissections. After the death of Joice Heth in 1836, Barnum arranged a public autopsy of her body at the New York City Saloon, to a large paying audience of medical professionals, reporters and the general public, ostensibly for the purpose of determining her true age. Staging a dissection as a form of public spectacle was not unheard of in the first half of the nineteenth century. The

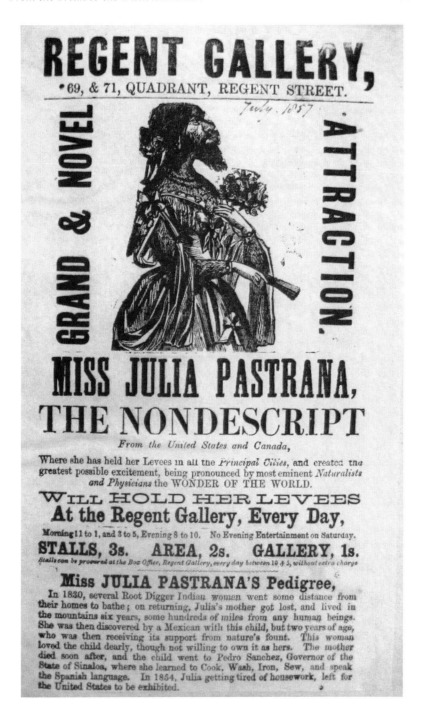

Fig. 12: Exhibition Handbill for Julia Pastrana at Regent Gallery (1857). Hope Collection, folio B.29. Courtesy of the Ashmolean Museum.

presiding anatomist at Heth's post-mortem, Dr Rogers, had previously participated in several high-profile autopsies that were simultaneously public spectacles and medical investigations. Notes Benjamin Reiss: "In 1830 he had dissected the body of the notorious pirate Charles Gibbs, and later turned over a 'Fac-simile of the Penis' of Gibbs to the Grand Anatomical Museum of New York" (36). The remains of another popular fairground exhibit, Saartjie Baartman, advertised as "the Hottentot Venus," were dissected by the French zoologist Georges Cuvier, who "presented to the scientific community a written report and her actual, excised genitals in a jar, and he apparently sent her skin back to England, where it was stuffed and put on display" (Reiss 130). Baartman's preserved body parts remained on public display at the Musée de l'Homme in Paris until 1974. Medical colleges engaged in similar practices, often exhibiting the dissected remains of high-profile subjects in their public museums: the Royal College of Surgeons in London undertook post-mortems on performers from commercial exhibition halls, such as Charles Byrne (the "Irish Giant") and Caroline Crachami (the "Sicilian Dwarf") whose remains are still on public display in the Hunterian Museum today. Similarly, the Royal College of Surgeons in Philadelphia dissected the original "Siamese twins" Chang and Eng, and the plaster cast made of their bodies is also on display in the Mütter Museum.[7]

The previous chapters of this book have drawn attention to the way public interest in spectacles featuring human bodies has intensified during those periods in which those bodies are undergoing rapid shifts in cultural status or significance. The invention of the "freak" body is another instance of this phenomenon, a construction brought about as a result of contemporaneous transformations in both medical and spectacular institutions, which gave rise to new ideas about bodies, their limits and their capacities. This chapter begins with a reconsideration of the history of the freak show found in recent critical studies, which argue that the increasing dominance of medical discourses over the twentieth century, which have come to redefine physical difference as disability, have brought an end to the freak show as a form of popular entertainment. This assumption, that spectacularised ways of seeing the body have been superseded by medicalised ones during the twentieth century, will be challenged in the second part of the chapter through a comparative analysis of the commercial photographs of professional freaks taken by Charles Eisenmann and those commissioned by the French neurologist Jean-Martin Charcot of his patients institutionalised for hysteria. The final section of the chapter extends current criticism of the freak show by examining its recent re-emergence as a form of popular entertainment, which has been driven to a significant degree by queer performers and those with disabilities. This section considers why, despite its (often well-deserved) reputation as the most disreputable of the public spectacles of human anatomy examined in this book, the freak show has proved such a fertile site for appropriation by those it has traditionally exploited and marginalised.

The Death(s) of the Freak Show

In recent years, as freak shows have emerged as a popular subject of academic histories and popular media alike, almost all writing on this topic has been characterised by its elegiac tone, lamenting the passing of a popular tradition that can be traced back well beyond the middle ages. Robert Bogdan's *Freak Show: Presenting Human Oddities for Amusement and Profit*, one of the first and most influential of these texts, is characteristic of what has subsequently become a critical consensus, casting its discussion of the history of the freak show in the past tense and concluding that "the splendor of the grand days of the freak show" have now come to an end (280).[8] Peter Schardt's more recent essay, in the aptly titled *The Last Sideshow*, is also framed as a eulogy, arguing that "there is no doubt now that the traditional Sideshow is disappearing" (5). Joe Nickell's *Secrets of the Sideshows*, from 2005, concludes by stating that "the century-long run of carnival sideshows […] is all but over" (345). "Since I interviewed these showmen," Nickell writes, "all have closed their shows" (349).

Recent fictional representations of the freak show share this nostalgic tone. Katherine Dunn's novel *Geek Love*, which follows the vacillating fortunes of "Binewski's Fabulon," begins as "the once flourishing carnival was fading" (7). Suffering from a shortage of freaks, the proprietor of Binewski's Fabulon, Aloysius Binewski, "decide[s] to breed his own freak show" (8) by administering a series of toxins and poisons to his wife. After she gives birth to a limbless "aqua boy," conjoined sisters, a telepath, and an albino hunchback dwarf, the show prospers for a time. This renaissance proves temporary, however, and the manufactured freaks prove so violently disruptive to their community that the institutional demise avoided at the start of the narrative is realised, definitively, by its end. More recently, Sarah Hall's novel *The Electric Michelangelo* focuses on the experiences of a tattoo artist working at Coney Island during the years of its waning popularity, during which the fairground was "sobering up from its early-century glory when even God had paid his entrance fee […]. Cy could sense the decline almost immediately after his arrival — the atmosphere was like coming late to a party where [...] the partygoers' eyes had begun to glaze" (184). By the end of the novel, the deterioration of the fairground has become palpable: "Coney Island looked sick to him.[...] Overnight it seemed as if the fairground had morphed from a potentially ugly thing into a hideous creature, a full-blown monster" (300).

Despite these recent obituaries, over the last decade freak shows have been flourishing and enjoying a period of renewed popularity. The reopening of the Coney Island Circus Sideshow, Sideshows by the Sea Shore, in 1982, paved the way for a new and rapidly expanding generation of troupes, including the Jim Rose Circus Sideshow, the Happy Sideshow, the Bindlestiff Family Cirkus, the Kamikaze Freakshow, Circus Contraption, Tokyo Shock Boys and Circus Amok, all of which draw heavily on the aesthetics and acts of the traditional ten-in-one sideshow.[9] This recent resurrection of the traditional freak show form, at the very moment a critical consensus has emerged about its imminent death, remains, however, largely unrecognised and under-examined in both popular histories and academic studies. Before turning to this recent reinvention of the freak show, this chapter begins with a re-examination of the claims about its death, recasting these not as empirical fact or evidence of the freak show's actual demise, but rather as a discursive construction which paradoxically enables its periodic reinvention as a cultural form. As Rachel Adams notes, the constant obituaries about the freak show might "say less about the freak show's demise than its restless plasticity.[...] Aware that scarcity or impending extinction are certain crowd pleasers, freak shows advertise not only the rarity of individual attractions, but the more general enterprise of human exhibition itself as a threatened practice" (211).

The death of the freak show is almost always attributed in the critical literature to the increasing medicalisation (and normalisation) of unusual anatomies over the twentieth century. This is said to have had two major effects on the freak show, both of them negative. Firstly, it is seen to have led to a radically decreased incidence, and even elimination, of some of the conditions that once made for popular sideshow exhibits.[10] As a result, laments the showman Ward Hall in *The Last Sideshow*, contemporary sideshows have entered a period of rapid decline, not because of a lack of public interest, but because of the decreased availability of unusual bodies for the sideshow stage: "Sideshows are more popular than ever," Hall insists, "the trouble is that there are not enough freaks" (Schneider 91).[11] This scarcity has produced a corollary change in the kinds of bodies available for display on the sideshow stage in the twentieth and twenty-first centuries, from a predominance of "born" freaks to "self-made" freaks. As Bogdan explains, these categories were used in traditional sideshows to distinguish between performers with anatomical anomalies present at birth and those who "acquired their physical oddity for the purposes of exhibition" (234). A survey of the kinds of performers featured at Coney Island and other contemporary sideshows over the last decade certainly reveals a significantly decreased incidence of congenitally non-normative bodies. In place of the "Siamese twins," "pinheads," "wildmen" and "armless wonders" exhibited at Coney Island's various ten-in-one shows in the late nineteenth and early twentieth centuries, recent seasons of Sideshows by the Seashore have instead featured such acts as Serpentina the snake-handler, Madame Twisto the contortionist, Eak the Geek and Diamond Donny the illusionist.

A second effect of twentieth-century medicine on the freak show is that it is understood to have taught the public to see anatomically unusual bodies in different ways, most significantly as the proper object of a medical rather than a public gaze, thereby rendering the spectacularisation of such bodies ethically problematic. As Bogdan argues, by the middle of the twentieth century, people with congenitally unusual bodies had been transformed in the cultural imagination from "freaks" or "oddities" to people with medical conditions requiring treatment: "the meaning of being different had changed in American society. Scientific medicine had undermined the mystery of certain forms of human variation.[...] People who were different had diseases and were now in the province of physicians, not the general public" (274). Rachel Adams reiterates this view in *Sideshow USA: Freaks and the American Cultural Imagination*: "Diagnosed in terms of recognisable pathologies, freaks lost the aura of mystery and wonder that once made them objects of visual fascination" (118). As a result, writes Rosemarie Garland Thomson in *Extraordinary Bodies: Figuring Physical Disability in American Culture and Literature*: "By 1940, the prodigious body had been completely absorbed into the discourse of medicine, and the freak shows were all but gone" (70).[12] Although "scientific and sideshow discourses had been entangled during the freak show era, they diverged towards opposite ends of a spectrum of prestige and authority as time went on," Garland Thomson writes, so that "scientists had transformed the freak into the medical specimen" (75).

This narrative of the freak show's history, which is based on the assumption that the determining context in which the anatomically unusual body is understood has shifted from that of the public spectacle to that of the professional clinic, is representative of the tendency that is evident throughout the history of popular anatomical exhibitions, in which the medical gaze is seen to supersede or correct popular ways of seeing bodies. The assumption underlying this tendency—that the medical and the spectacular represent distinct and separate ways of seeing—is evident not only in recent histories of the freak show but also in the critical accounts of one of the freak show's most direct historical precedents: the public exhibitions of "human monsters" in early modern Europe, whose eventual (and temporary) disappearance is also attributed to the rise of scientific and medical cultures. Accordingly, it is instructive to briefly consider this period, as it provides both a historical context for the public exhibition of unusual anatomies within which to situate the later freak show, while also complicating common critical assumptions about it.

In the sixteenth and seventeenth centuries, as in the nineteenth, the public was fascinated by commercial exhibitions of unusual bodies, which were held in public places such as markets and fairs; and in the sixteenth and seventeenth centuries, as in the nineteenth, interest in these exhibitions was catered to by the publication of texts for a popular audience, including illustrated broadsheets with evocative titles such as *A Certaine Relation of the Hog-Faced Gentlewoman called Mistress Tannakin Skinker* (Skinker n. pag.), or *A Monstrous Shape. Or a Shapelesse Monster. A Description of a female creature borne in Holland … now to be seen in London* (Price n. pag.), as well as those for a more educated audience, including philosophical and political texts like Montaigne's "Of Cripples" or Hobbes' *Leviathan*.[13] Historians of the early modern period, like those of contemporary popular culture, have tended to attribute the eventual waning of public interest in such exhibitions and representations of monstrous bodies to the increasing influence of a modernising medical gaze, which is widely seen to have acted as a scientific corrective to a more spectacular and credulous one. Colin Clair succinctly articulates this argument when he contends that the "spread of education and greater medical knowledge" during the Enlightenment led to an inevitable waning of interest in "such 'marvels' as the woman with three breasts, the pig-faced woman, and the ubiquitous mermaid" (93).

There is, however, another way to understand the reconfiguration between the medical and the spectacular gaze as reflected in the early modern exhibitory culture. Although the rise of modern medicine is said to have put an end to early modern spectacles of unusual anatomies, by positioning the medical and the spectacular as newly separate and opposed representational economies, closer examination reveals that the two ways of seeing emerged and developed simultaneously, mutually informing and transforming one another in a way that problematises attempts to situate them as two distinct forms of knowledge.

We can consider this, and the significance of this history for the way the freak show has been understood in contemporary scholarship, by turning briefly to two landmark texts published during this time.

Ambroise Paré's *Des monstres et prodiges* (1585) was perhaps the first early modern text to undertake an anatomical study of bodily difference.[14] Paré was a physician and surgeon, and although his text reproduces many of the fanciful and folkloric accounts of the causes of bodily "monstrosity" that were so popular in his day — and were the subject of countless contemporaneous books — *Des monstres et prodiges* differs from these in incorporating anatomical studies of physical difference and attempting to account for its incidence through consideration of the medical and environmental factors by which it was produced. Thus, although *Des monstres et prodiges* takes seriously, for instance, the role of maternal impression in producing monstrous bodies (according to which theory a stubborn woman might give birth to pig-headed offspring, for instance), Paré also recognised that monstrous births could be caused by injury during pregnancy and other disturbances to foetal development.[15] This medicalised approach to unusual embodiment, and its concomitant redefinition of monstrosity as anatomical difference, reflects two important shifts in early modern understandings of the body. Firstly, as Norman Smith argues, images of the monstrous body in the sixteenth century represent a shift away from a medieval fascination with distant monstrous races towards a marked interest in "the monsters [the public] could see about them — anomalous births, strange events, occurrences contrary to nature" (267). In other words, whereas monsters in the middle ages were imagined as elsewhere, during the early modern period monsters moved from the margins of culture to its centre. Secondly, and co-extensively, the early modern fascination with monstrous bodies and births constituted a reimagining of what the monster was: no longer seen as a religious omen to be deciphered, by the late sixteenth century the monster was increasingly reconceptualised as secular and naturally occurring: something to be explained by doctors rather than priests.

Such epochal transformations in the meaning of monstrosity at this time would seem to attest rather to its heightened cultural significance and popular interest than to its decline, an impression reinforced by the publication, a century later, of John Bulwer's *Anthropometamorphosis: Man Transform'd; or, the Artificial Changeling*. Bulwer, a medical practitioner who had previously written texts on deafness and gesture, described his study as a work of "Corporall Philosophy" ("A Hint of the Use of this Treatise" n. pag.) which he intended as a denunciation of practices of self-transformation.[16] Providing the first cross-cultural history of body modification, Bulwer's text, like Paré's, is replete with the apocryphal tales of exoticised others that were so popular at the time, such as his account of the inhabitants of Fanesii, "whose Eares are dilated to so effuse a magnitude, that they cover the rest of their Bodies with them, and have no other Cloathing" (*Anthropometamorphosis* 141). In this respect, and in another parallel to Paré's text, although Bulwer's work may appear merely an exemplar of a popular genre of his day—compendia of tales about exotic and ancient races excerpted from classical texts—*Anthropometamorphosis* is nevertheless a landmark publication in two ways. Firstly, it reflects an epochal shift in the dominant mode of thinking about bodily change. Whereas Paré's text understood transformation as something that happened spontaneously and unpredictably *to* people, Bulwer's focus is on acts of intentional modification, on practices of self-making that are deliberately undertaken by the subject him/herself. Secondly, *Anthropometamorphosis* is the first text to incorporate illustrated descriptions of contemporary English fashions alongside those of historically or geographically distant cultures. Bulwer himself was trenchantly against such practices. Writing during a period marked by regicide and civil war, in which the relationship between church and state was being rapidly transformed, Bulwer, a monarchist, compared the self-made body to the secular state. He saw both as monstrous constructions that deformed the natural, divinely shaped order of the world.[17]

The accounts by Paré and Bulwer of the monstrous body both make clear the extent to which this was simultaneously understood through the new medical discourses and knowledge emergent in the sixteenth and seventeenth centuries and through older, more spectacular ways of seeing such bodies, transforming these but also a product of them. Moreover, given that the period between the publication of these two texts marks the high point of the early modern fascination with the monstrous body, it might be argued that the development of medical ways of seeing the body catalysed public interest in such displays rather than bringing about their end, generating a heightened curiosity in this exhibitory form. As Margrit Shildrick argues in her study of the monstrous body from the early modern period to the present: "What seems to be a simple narrative of progressively more rational approaches to the issue of monstrous form obscures a far more complex process of contestation in which a whole range of modernist parameters of knowledge—truth and fiction, self and other, normal and abnormal—are at stake" (*Embodying the Monster* 27). Considered from this perspective, medical discourses can be seen not to have brought an end to interest in the figure of the monster; rather, the monster becomes the site at which to make sense of the changes to concepts of the body produced by these new medical discourses. As is evident in the texts of Paré and Bulwer, medical knowledge about the body is itself shaped by existing ways of seeing and by popular thought: monsters in the early modern period might be seen to become objects of such fascination precisely because they provide a space in which these mutual transformations can be examined and debated.

Taking the history of the monster into account allows us to reconsider that of the freak, too, not as one in which a medical way of seeing and understanding that body comes to supplant an earlier and more credulous one, but rather as one in which the relationship between the medical and the spectacular is actively and mutually reconfigured. Thus the figure of the freak, like that of the monster before it, cannot be seen to have an essential meaning in and of itself but, rather, functions as a highly flexible category, a stage on which various ideas and concerns about the body are played out, and around which new ways of seeing are simultaneously constructed and contested. In her analysis of gothic fiction, Judith Halberstam argues that the monster is a "meaning machine" (*Skin Shows* 21), one that produces "interpretive mayhem" (2): "The monster always represents the disruption of categories, the destruction of boundaries, and the presence of impurities" (27). For Margrit Shildrick, similarly, the monster is a force of destabilisation that cannot be safely quarantined: "time and again the monstrous cannot be confined to the place of the other; it is not simply alien, but always arouses the contradictory responses of denial *and* recognition, disgust *and* empathy, exclusion *and* identification" (*Embodying the Monster* 17; original emphasis).

The figure of the freak, too, is often closely associated with the problematisation of categories in a way that raises questions about the significance of those bodies to which it offers no clear answers. As Tromp and Valerius recognise in their recent overview of the British Victorian freak show:

> Freak shows attracted audiences by inviting the public to engage in epistemological speculation. Was the Feejee Mermaid a fake? Was the bearded lady really a man? Audiences paid for the opportunity to take a look and decide for themselves. Significantly, this interrogatory practice made freak shows volatile interpretive spaces that repeatedly called the boundary between the imaginary and the real into question, and by extension challenged the authority of discourses like medical science to name and explain the significance of the human body. (Tromp 8)

This emphasis on the uncertainty of the meaning of the freak body draws attention to the limitations of medical discourses' ability to determine or circumscribe its cultural significance. In this way, the history of the freak show can be (re)read not as one in which the relationship between the medical and the spectacular, the normal and the freakish, the bourgeois public sphere and the chaotic fairground, were gradually stabilised into separate fields (in which one is increasingly dominant and the other increasingly marginal), but, rather, as one of ongoing interdependence and mutual reconfiguration between these apparently separate and opposed fields.

This is reflected in the way the space of the freak show stage has been, since its inception, a site of explicit debate and critical reflection about the meaning of the term "freak" and its impact on those to whom it is applied. A notable instance of this occurred in 1899, when the performers in the Barnum and Bailey Circus's "Freak Department" staged a protest against the use of the word "freak" in circus advertising and on banners (Lentz 26).[18] The manager of the Freak Department was subsequently required to remove the term from publicity material for the sideshow. This moment is significant for a number of reasons, not least of which is that it marks one of the first recorded occasions on which those who are the objects of such spectacles voice an opinion about the terms under which their bodies are exhibited (a point whose significance we will return to in the final section of this chapter). At the same time, it also demonstrates that debates about the significance of the term "freak" are as old as the freak show itself.

Despite the objections of the Barnum and Bailey Circus performers at the end of the nineteenth century, however, the term "freak" continued to be widely used within sideshows until well into the late twentieth century. In 1984, nearly a century after the Barnum and Bailey Circus performers' revolt, the Sutton Side Show at the New York State Fair was also criticised for its use of this term to describe its performers. On this occasion the complaint came from outside the sideshow itself, from a disability rights activist. As a result, the sideshow was relocated from the prominent midway area to the back of the fair, drastically reducing its audience and profitability. One of the performers, Otis Jordan ("The Frog Man"), complained about the prohibition on his performance being advertised as part of a freak show, and the motivating assumption that he was being exploited (Bogdan 279–81).[19] A hundred years after the protest at the Barnum and Bailey Circus, then, Jordan was actively arguing *for* his right to allow himself to be marketed as a freak. Jordan's ability to advocate for the right to advertise himself as a freak is, it should be recognised, a product of this earlier insistence by professional freak performers that they be heard rather than simply stared at, which paved the way for later generations of performers to exert the same right, even those who, like Jordan, would use that voice to express an opposing point of view.

The debate provoked by the Sutton Side Show's use of the term "freak" has rightly been identified as a pivotal moment in the history of the freak show, because it exemplifies all the concerns surrounding the significance of the word "freak," its function within the space of the spectacle, and its consequences for those exhibited as such. This case is particularly important because it is one in which the disabled performer and the disability advocate take opposing stands. For Bogdan, who interviewed Jordan after his departure from the Sutton Side Show, this case represents the moment in which the subjects who have been widely assumed to be the passive victims of the freak show culture first exert their right to exhibit their bodies for profit if they so choose. Bogdan writes:

> According to him, [joining the side show] was the best thing that ever happened. He likes to travel and meet new people and his new pro-fession enabled him to buy a small house back home which he lives in when the show winters. He has no complaints except one. He thought the woman complaining about his being exploited ought to talk to *him* about it. He would tell her there "wasn't anybody forcing him to do anything." As he put it, "I can't understand it. How can she say I'm being taken advantage of? Hell, what does she want for me—to be on welfare?" (280)

Dick Zigun, founder of Coney Island's Sideshow by the Seashore, who subsequently employed Jordan as "The Human Cigarette Factory," also challenged the Sutton Side Show's decision: "Who's exploitative, the critic who condemns the performer or the producer who pays him a salary?" (quoted in Adams, *Sideshow USA* 216).

David Gerber, however, takes issue with what he sees as Bogdan's idealistic interpretation of Jordan's situation. Gerber questions the validity of Bogdan's argument that Jordan both consented to and actively agitated for his right to perform under the label "freak." Gerber argues that meaningful and informed consent can only be given in situations in which subjects have a range of choices available to them: "But if Jordan's choices in life have been reduced to participating in a freak show or 'being on welfare,' it really does not appear that he has had much choice at all" (49). These do indeed appear to have been Jordan's only options: to accept a role in the public space of the freak show as "The Frog Man"; or to accept the medico-legal definition of his body as disabled, and go on welfare. As this "choice" makes clear, the difference in meaning between these two terms also has important consequences for the spaces in which Jordan's body can subsequently move. Whereas the word "freak" is the name of a theatrical persona, a public performer, who is highly visible on the public stage, the word "disabled" relocates that body in the private or professional sphere. Thus the change in nomenclature from a "freak" to a "disabled person" has a profound impact not only on the conceptualisation but also on the visibility of the body to which it is applied and the spaces to which that body has access. Such a shift can be seen to underlie Gerber's own discomfort; the emergence of disability as a category has made it conceptually unsettling for contemporary audiences to see congenitally unusual bodies spectacularised on the public stage—it is now difficult to see such bodies as anything other than exploited or unfortunate. For Jordan, it would appear, this way of seeing and knowing his body was even more objectionable than the compromises entailed by performing as a professional freak. But it also indicates the enduring appeal of the freak show in the twenty-first century: in a time in which the meaning of such bodies is the object of so much uncertainty and debate, the freak show provides an occasion to stare at the very bodies that most members of contemporary audiences have been trained since child-hood *not* to stare at.

What Jordan's case makes clear is that, while questions about the ethical treatment of those with unusual anatomies remain a necessar-ily and importantly central part of public receptions to the freak show, there are no simple answers to how these concerns should inform its exhibitory styles and institutional practices. Rather than simply condemn the traditional freak show as an unethical and exploitative space—although it certainly could be and often was exactly that—we might see this as a space in which questions about the ethical treatment of bodies and bodily difference were, and continue to be, raised and debated. Indeed, the inevitability with which such questions are raised is itself significant, reflecting the degree to which ethical considera-tions define the conceptual and cultural field in which freak shows are currently understood, and in which the intersection between medical and spectacular cultures are presently negotiated. It is for this reason that this chapter is less concerned to make a determination about the ethics of the freak show than to consider how freak shows have been discursively and institutionally shaped by changing assumptions about what constitutes ethical relations, and what this tells us about the changing status of unusual anatomies exhibited in the public sphere.

The inevitability with which discussions about the freak show come to focus on their ethical considerations demonstrates how central but unresolved questions about the meaning of the bodies put on display, and about the legitimacy of that display itself, remain to this history, reflecting the long-standing cultural function of exhibitions of anomalous bodies as "a locus of production, the site of contested meaning" (Shildrick, *Embodying the Monster* 10). The instability and undecidability associated with the figure of the freak, the intensity of the debates that continue to swirl around it, are important because they indicate that medical discourses have not come to uniformly or unequivocally dominate our understanding of the meaning of such bodies; on the contrary, the space of their public exhibition remains one of active debate and ongoing negotiation. In the context of the critical consensus that the history of the freak show has been shaped (and ended) by the progressive dominance of medical discourses, it is thus salutary to remember how elusive the meaning of the freak body has remained, how difficult to control or stabilise, as we will see in the following section.

The Clinical and the Spectacular Gaze: Photography in the Popular and the Professional Spheres

Unlike the forms of exhibition examined in the previous chapters, freak shows left behind a vast archive of material, much of it in pictorial form. The main reason for this is that freak shows developed at the same time as photography, and made extensive use of this new technology for advertising purposes. In a period gripped by what William Darrah has called "cartomania"—a frenzy for collecting the cards produced by professional photographers in the years before affordable personal cameras were available (4–11)—commercial photographs of freak performers enjoyed a wide and popular circulation.[20] Medicine also made extensive use of photography at this time to record images of wounded bodies or pathological conditions, incorporating these within case studies that were circulated primarily among a professional audience. While their shared early uptake of photography is another manifestation of the close association and historical independence of medical and spectacular ways of seeing bodies, at the same time, and tellingly, the status accorded to such images in these contexts, the sorts of representational economies of which they were a part, were very different. Medical uses of photography were predicated on a widespread belief that photography provided a purely mechanical visual record of its object, an unmediated representation of the thing itself. Commercial photographs of freak performers, on the other hand, were highly theatricalised, drawing heavily on the conventions for representing anatomically unusual bodies established in and by contemporaneous sites of public spectacle.

This section of the chapter will examine the dis/continuities between medical and theatrical ways of seeing the body in the late nineteenth century through a comparative analysis of the freak portraits taken by Charles Eisenmann and the photographs of hysterics commissioned by the French neurologist Jean-Martin Charcot. The conventions these sets of photographs reproduce have much to tell us, not only about the changing significance of the bodies they depict, but about the changing status of the knowledge and interpretative frameworks by which this significance was (trans)formed in the late nineteenth century. As we have seen in the first section of this chapter, the freak body is more a site of destabilisation and uncertainty than a figure of physical difference whose meaning is fixed. Lilian Craton recognises that freak shows "resist a unified presentation of the meaning of physical difference, but they do show difference to be meaningful" (4). It is just this construction of physical difference as an important site of meaning-making, but whose significance remains both unsettled and unsettling, that we see played out in the photographs of Eisenmann and Charcot.

Charles Eisenmann is widely recognised as the foremost photographer of freak performers.[21] His photographs provide an extensive archive of images of fin de siècle professional freaks, from the most successful and well-known performers to those whose only remaining historical record is Eisenmann's portraits of them. Eisenmann's photographs include now-iconic portraits of Theodor Jeftichew ("Jo-Jo the Dog-Faced Boy"), William Henry Jackson (Zip, the "What is it?"), and a family portrait of Chang and Eng (the original "Siamese Twins"). Other of Eisenmann's subjects included Madame Devere (the "Bearded Lady"), Anna Leake Thomson (the "Armless Lady"), Myrtle Corbin (the "Four-Legged Girl from Texas") and Fanny Mills (the "Ohio Big-Foot Girl"). Eisenmann's portrait of Fanny Mills is, in many ways, exemplary of his photographic treatment of freak performers [Fig. 13].[22] The composition of this image, taken for advertising and promotional purposes, is representative of the style of these commercial *cartes de visite*, reproducing many of the theatrical conventions by which bodies were constructed as those of professional freaks in and through their public display.[23] The image included here is one of a series of portraits Eisenmann took of Mills, in which her dress is depicted at first entirely covering her legs, and then raised progressively higher in each image, together constituting a sort of strip tease of freakery.

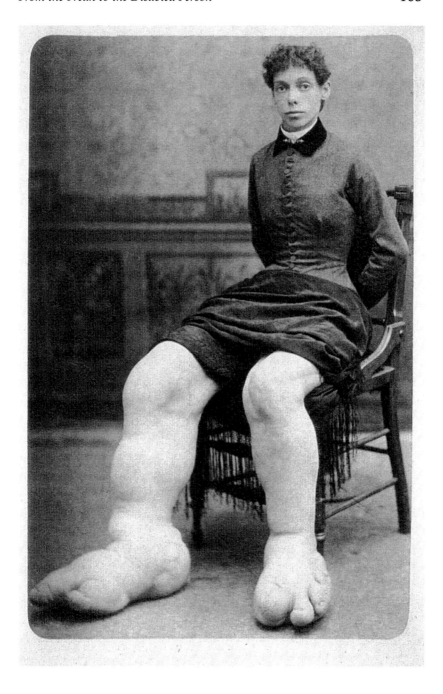

Fig. 13: Portrait of Fanny Mills, Charles Eisenmann (undated). In *Monsters: Human Freaks in America's Gilded Age: The Photographs of Chas Eisenmann.* Courtesy of Michael Mitchell.

Hence the photographic gaze is here recognisably informed by the spectacular gaze of the sideshow, and particularly by its techniques for exaggerating corporeal differences in which, for instance, "dwarfs" would be exhibited alongside "giants," "fat ladies" alongside "human skeletons," while "bearded ladies" would be garbed in hyper-feminine frilly dresses, and so on. Similarly, those whose physiognomies could be constructed as racially other would be photographed in exoticised settings (like the San Salvadorean microcephalics, Maximo and Bartolo, exhibited as "The Last of the Aztecs"), while white subjects with unusual bodies like Fanny Mills were most often photographed in familiar domestic contexts. In Mills' portrait, the contrast between the exceptionality of her body and the normalcy of the drawing room in which she is placed is used to heighten the visual impact of the extraordinary dimensions of her body. The angle of Eisenmann's photograph, which is shot slightly upwards, further emphasises the disparity in proportion between her small upper body and the excess and irregularity of her lower limbs. The corseted dress, with its raised skirt, accentuates not only the disjunction between her neatly contained torso and the spreading largeness of her feet, but also constitutes a photographic translation of her public exhibition and its spectacularising revelation of Mills' physical anomaly.

The significance of Mills' body for nineteenth-century audiences, and the interest it provoked (reflected in the existence of a market for portraits like this), is not simply a product of the new technology of photography, of course, or the extent to which it faithfully reproduced the conventions of popular spectacle. Rather, the photographic archive that includes images like Mills' helps us to recognise what it was about certain kinds of embodiment that made them such a fascinating but frightening spectacle for late nineteenth-century audiences. We have seen in each of the previous chapters that popular medical discourses and public spectacles from the eighteenth century onwards were united in their representations of the body and its health as a matter of personal responsibility and self-control, something each individual had a personal as well as a social duty to care for and cultivate. Within the context of American culture in the nineteenth century, during the period in which freak shows flourished as a form of popular entertainment, this attitude towards the body was greatly intensified. The "ideology of liberal individualism" that prevailed in the USA at this time required a particular kind of body, Rosemarie Garland Thomson recognises, one that "is a stable, neutral instrument of the individual will" (*Extraordinary Bodies* 42). In this context, bodies we would now refer to as "disabled" were both frightening and fascinating for able-bodied spectators because they trouble this "American ideal":

> Disability's indisputably random and unpredictable character translates as appalling disorder and persistent menace in a social order predicated on self-government ... the disabled body stands for the self gone out of control, individualism run rampant: it mocks the notion of the body as compliant instrument of the limitless will and appears in the cultural imagination as ungovernable, recalcitrant, flaunting its difference as if to refute the fantasy of sameness implicit in the notion of equality. (43)

A body like Mills', whose dimensions are so clearly beyond any measure of self-control to contain, represents the feared flipside of the self-making bourgeois subject: like the liquid body of the spermatorrhoeaic spectacularised in the contemporaneous museums of anatomy "for men only," Mills' body was an object of fascination for nineteenth-century audiences precisely because it made visible the fears about the unruly or uncontrollable body that haunt the docile and well-disciplined subject.

As Garland Thomson recognises, such bodies "are made to signify what the rest of Americans fear they will become. Freighted with anxieties about loss of control and autonomy that the American ideal repudiates, 'the disabled' become a threatening presence, seemingly compromised by the particularities and limitations of their own bodies" (*Extraordinary Bodies* 41). The significance of Eisenmann's photograph of Mills is that it at once reflects this anxiety, as it circulated within a nineteenth-century exhibitory culture, and holds out the promise of its containment. Thus, it demonstrates how the emergence of photography extended the existing exhibitory culture of its time while also transforming it, in ways that again draw our attention to the mutual interdependence between spectacular and medical discourses, between technologies of the image and technologies of the body. Just as the museum transformed objects into exhibits through the process of labelling them and putting them on display, explaining their significance and arranging them in an orderly fashion within the larger systems of which they were a part, so does Eisenmann's photograph transform Mills' body into the exemplar of a particular type, complete with its own explanatory label—in this case, the medical biography included overleaf on her *carte de visite*.[24] Like the label appended to the museum exhibit, the photograph and narrative of Mills' body promise us certainty about its meaning, a meaning which appears to have been safely fixed and stabilised through the recording of its image and medical condition.

A similar drive underlies the medical photographs commissioned by Jean-Martin Charcot (1825–1893) in the course of his work with hysterics at the Hôpital Pitié-Salpêtrière. For Charcot, working within the context of a professional medical facility, photography was a useful diagnostic tool, providing a visual illustration of his cases for his medical records and playing a pivotal role in the identification of the symptoms of hysteria. Although Charcot's practice drew on the work of earlier practitioners such as Landouzy, Brachet and Briquet,[25] it is Charcot who is widely credited with recognising hysteria as a treatable condition with an identifiable set of symptoms, thereby imposing "a persuasive set of laws" on what had until then been regarded as the "anarchic shapelessness and multiple symptoms—paralyses, muscle contractures, convulsions, and somnambulism—of hysteria" (Showalter 33). As Sander Gilman argues, photography was a crucial part of Charcot's research into and treatment of hysterical patients. His photographs were designed "to capture hysteria's stages and processes as they represented themselves on the visible surface of the patient, on the patient's physiognomy, posture, actions, as a means of cataloguing the disease process" ("The Image of the Hysteric" 352). Thus it was through its photographic documentation that hysteria came to be seen no longer as the mysterious expression of a fundamentally unknowable or untreatable condition, but as a recognisable condition with a legible set of symptoms that could be read on the body of the patient.

While Charcot's photographs were thus taken and intended for medical purposes, they nonetheless clearly functioned within the same larger exhibitory culture as did Eisenmann's. As Georges Didi-Huberman notes in *Invention of Hysteria: Charcot and the Photographic Iconography of the Salpêtrière*, Charcot himself understood la Salpêtrière as "an anatomo-pathological museum," and explicitly reconfigured the hospital as an exhibitory space (30).[26] Large images of his patients were hung in the theatre where he delivered his series of public lectures, while their photographs were published in *Iconographie photographique de la Salpêtrière* and made available for public sale (Showalter 31). Moreover, despite his claims that they simply recorded the hitherto mysterious symptoms of hysteria, closer examination of Charcot's photographs demonstrates that medical knowledge about the body often constructs or transforms what it claims to discover or document. In this way, like Eisenmann's photographs, Charcot's photographs of hysterics have a theatrical dimension, within which hysteria is constructed as an object of spectacle as well as medical study.

We can see this process of medical spectacularisation in play by turning to one of Charcot's best-known photographs: the series of "attitudes passionnelles," featuring his patient Augustine, which were published in *Iconographie photographique de la Salpêtrière*. In the image entitled "Extase" [Fig. 14], Augustine is represented as caught in the midst of the convulsive spasm that was, for Charcot, the "central sign of hysterical disorders" (Showalter 33). Although rendered static by the medium of the photograph, the movement of Augustine's body is dramatised by the falling of her hospital gown from one shoulder, which partially reveals her breast. Like Mills, she is depicted in an apparently domestic setting (a bed) that is in fact the product of photographic composition: Mills in the faux-drawing room of Eisenmann's studio, and Augustine within the institutional context of the hospital. The composition, backdrop and angle of this photograph all serve to spectacularise—if not sexualise—Augustine's body: the reach of her open arms is paralleled in the open space between her legs, positioned at the foreground of the image and level with the camera, behind which the viewer can see the disarranged pillows and sheets of the bed.

In this respect, Charcot's photograph of Augustine, like Eisenmann's portrait of Fanny Mills, is composed according to a set of conventional assumptions about a physical condition of which Augustine is taken to be the embodiment. Hysteria has a long and close association with female sexuality, as its etymological origins in the Greek *hystera*, or womb, indicates. The hysteric has often been represented, in Nicole Edelman's words, as "a lascivious, erotic and rebellious figure" (7), the very embodiment of an unruly sexuality. This is clearly reflected in Charcot's photographs of Augustine and his depiction of her lack of control over her spasmodic bodily movement. Much has been written about the ways in which hysterics challenged the normative assumptions about femininity of the late nineteenth century; what Charcot's photograph makes clear (as distinct from his incarceration or treatment of these women as patients) is the way photography itself was one of the technologies by which medical institutions sought to contain the threat of these unruly and uncontrollable bodies.

In this respect, Charcot's photographs, like Eisenmann's, need to be understood as manifestations of the wider encyclopaedic and archival tendencies that were so central to the nineteenth-century disciplinary society. As Allan Sekula recognises, police archives at this time made extensive use of photographic portraits (along with other forms of bodily imaging, such as fingerprints) as a means by which to identify particular criminal types. During the second half of the nineteenth century, photographs were commissioned on a wide range of subjects—primarily those who, like freaks or hysterics, were identified as in some way anomalous or deviant: prostitutes, criminals, homosexuals, the insane, chronic masturbators, people with deformities and those with intersex conditions. As Dana Seitler writes: "it

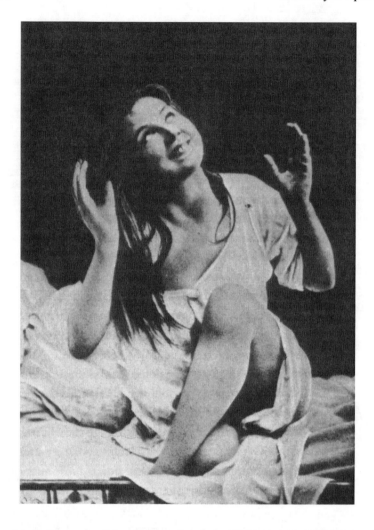

Fig. 14: Photograph of Augustine taken at la Salpêtrière by Paul Régnard. In *Iconographie photographique de la Salpêtrière*, vol. II, plate xxiii. Courtesy of the Boston Medical Library in the Francis A. Countway Library of Medicine.

was the medium of the photograph that was most often deployed within nineteenth and early-twentieth century science, medicine, and state institutions as an instrument to survey, record and account for the human body" (82). Just as the general public was seized by a wave of cartomania, then, so were these public and professional institutions gripped by an archive fever, a new archival imaginary within which all bodies would be both individually identifiable and categorisable by general type.[27] Museums and other exhibitory spaces reflected a similar tendency, writes Beth Lord: "like encyclopaedias and libraries, museums are monuments of the eighteenth-century drive to categorise, classify, and order the world into a totality universal in scope and universally intelligible" (2).[28]

At the same time, however, and as Charcot's photographs show, while photography was an integral part of the nineteenth-century disciplinary society, it also served to transform disciplinary institutions into exhibitory spaces. This transformation is, moreover, one that serves to destabilise the central purpose of these disciplinary institutions, problematising the very categories and classifications they were intended to identify. As Foucault famously argues in volume one of *The History of Sexuality*, the Victorian period was characterised not by the repression of sexuality but, on the contrary, by the invention of contemporary concepts of sexuality through the proliferation of discursive formations and institutional structures designed to identify, categorise and regulate bodily practices, stabilising these into an ordered series of identities. The focus of this activity was, as it was for the photographic archivists of the same period, or the promoters of freak shows, on the anomalous or marginal: "a whole web of discourses, special knowledges, analyses, and injunctions settled upon it," Foucault writes, producing not the "exclusion of these thousand aberrant sexualities, but the specification, the regional solidification of each one of them" (44). At the same time, however, and as Foucault is well aware, it is through the process of their cataloguing that these types are produced: they are the result, not the object, of their identification and documentation. Despite their promise to contain the unruly body, to provide a fixed image and identity of it, what the archival tendencies of nineteenth-century disciplinary institutions reveal is, on the contrary, precisely their failure to control and determine the meaning of these bodies.

We can see this by returning to Mills' *carte de visite*, in which the two elements that constitute it—the photographic portrait and the medical description—are mutually destabilising as much as mutually reinforcing. While the purpose of the medical narratives included on *cartes de visite* was clearly to provide an interpretive framework through which these bodies could be read, reducing their meaning to their identified medical condition, the increasingly medicalised view of bodily difference these images (re)produced was also a significant contributing factor in the demise of the exhibition of professional freaks over this very period, redefining this body as disabled in a way that moved it off the public stage and into a private or clinical setting. In this way, Eisenmann's photographs can be seen to trace and contribute to the decline of the very phenomenon they document: rather than stabilise the identities of the bodies he photographed, Eisenmann's portraits in fact record their transformation and subsequent vanishing from the public sphere.

A similar trajectory can be identified in Charcot's photographs of hysterics like Augustine. These photographs are commonly said to have enabled the identification of the fixed symptoms of hysteria, and thus to have "solidified" (to use Foucault's term) its functionality as a medical category. Yet at the same time, and in keeping with the dynamics we have seen in Eisenmann's photographs of professional freaks, the performances of the hysteria captured in these images actually (and paradoxically) did much to contribute to the decline of hysteria itself as a diagnostic category over this period—particularly, as Michael Roth notes, by generating criticism within the medical community that these were too staged, and the women in them simply "acting out" symptoms at their physicians' behest (23). In consequence, as Showalter has shown, while diagnoses of hysteria reached epidemic proportions in the late 1800s, by the early twentieth century hysteria had all but disappeared as a diagnostic category (Showalter 17). Thus, just as Eisenmann was taking his portraits of professional freaks at the very moment the idea of the freak body was in the process of being redefined and relocated away from the public gaze, so were Charcot's photographs taken in the period immediately prior to the decline of hysteria as a common medical diagnosis. What we see, then, in the photographs of Charcot and Eisenmann is not the identification and stabilisation of categories of non-normative bodies through the recording of their images, but, on the contrary, the elusive moment of their transformation and disappearance. Hence, these images reveal not the unidirectional construction of non-normative bodies by oppressive systems of power/knowledge but, more ambiguously, the complex interrelationships between bodies, medical knowledge, spectacular representations and institutional power that characterised fin de siècle exhibitions of anatomical difference.

As Dana Seitler argues in "Queer Physiognomies; Or, How Many Ways Can We Do the History of Sexuality?", while fin de siècle medical and scientific photography provides an especially fruitful medium through which to trace the emergence of modern concepts of (homo) sexuality, this material simultaneously problematises the assumption that homosexuality is a stable category, because photographs of "the sexual degenerate, homosexual or pervert existed indiscriminately among other examples of perceived degeneracy and deviance from this period" (84). Like the images of performing freaks and medicalised hysterics examined here, Seitler argues that photographs of homosexual bodies at the fin de siècle reveal a discursive and institutional attempt to identify and stabilise constructions of a "perverse" body that remained, nonetheless, fundamentally plural and elusive. In a conclusion strikingly pertinent to the photographs taken by both Eisenmann and Charcot, Seitler writes:

> [t]he desire to corporealize sex is an attempt to produce, in visible form, a noncontradictory sign that would organise and align a series of sexual practices, social behaviours, and medical etiologies within a readable image [... T]he unstable proliferation of the definitions and embodiments of perversity ultimately render the perverse body a hybrid and indeterminate one, formulating the variables of perversion into an unmanageable figure—a multiply produced, polyvalently diseased, indistinct image. (97)

In a similar way, although the stated purpose of the photographs of hysterics Charcot commissioned was to record and catalogue the symptoms of the disease they supposedly document, their meaning, like those of Eisenmann's freak photographs, is not fixed and stable but, on the contrary, in the process of constantly transforming—transformations, moreover, to which his photographs actively contributed. This is significant because it problematises one of the central assumptions underpinning criticism of the freak show in particular and of the public exhibition of unusual anatomies more generally: that this is a history in which medical discourses have become progressively more dominant, eventually coming to overwrite popular ways of seeing such as that of the freak show, and replacing these with new scientific categories like that of disability. As the photographs of Charcot and Eisenmann show, neither medical nor sideshow discourses succeed in fixing or determining the meaning of these anomalous bodies, but continue to work both with and against one another. It is this instability at the heart of the freak show, this incapacity to be contained within either the exploitative space of commercialisation or the pathologising space of medicalisation, that has allowed it to outlast the repeated claims about its imminent demise, and which gives it such potential as a site of reappropriation and resignification, a potential of which contemporary freak performers have taken full advantage.

Coney Island and the Postmodern Sideshow

As we saw in the first section of this chapter, despite the repeated pronouncements of its death, the freak show has not disappeared as a form of popular entertainment in the twenty-first century but, on the contrary, is currently experiencing a strong resurgence. Coney Island's Sideshows by the Seashore, as the first of a new generation of freak shows, has played a vital role in reinvigorating this culture and providing an important training and performance space for both emerging and established performers. Opened in 1982 on the site of David Rosen's Wonderland Sideshow, Sideshows by the Seashore was established by the non-profit performing arts and historical corporation, Coney Island USA, founded by Dick Zigun, with a mission to "document, preserve and further the unique arts" of the sideshow in a way that both "interprets the past and experiments with the future" (Coney Island website). Funded by the National Endowment for the Arts, the New York State Council for the Arts and the New York City Department of Cultural Affairs (Siegel 107), the Coney Island Circus Sideshow is thus simultaneously a celebration of its freak show heritage, a museum dedicated to that heritage, and a postmodern reinvention of it.

Visitors to Coney Island today would, however, be forgiven for assuming that Sideshows by the Seashore is a relic of the amusement parks' early twentieth-century heyday, like the iconic 271 ft 1941 Parachute Jump and original roller coasters from the 1920s that are landmarks of its midway. The sideshow is housed in a theatre space whose aesthetics reinforce this impression, circled by the distinctive orange and yellow banner art of early twentieth-century freak shows, featuring highly stylised portraits of acts such as Madame Twisto, Serpentina, Eak the Geek and Insectavora [Fig. 15].[29] Most striking, however, is the prevalent use of the term "freak", which, as Michael Chemers notes, represents a deliberate attempt to reclaim a term that remains widely perceived as a pejorative (7–8). This word features prominently in the Sideshow's publicity, painted on the awning between the banner panels for each act, appearing in the large sign advertising the "Freak Bar," and repeated in the neon "freaks" sign hung in the theatre entrance way. The sideshow also hosts regular series of special events such as the Girlie Freakshow and the Superfreak Weekend. The show itself also reproduces the style and acts of a traditional ten-in-one sideshow, with an outside talker and a ballyhoo platform (a small stage in front of the theatre) to tout for business. Currently, its acts include: a sword swallower; a snake handler; an illusionist; a bed of nails act, in which members of the audience are invited to stand on the performer's chest; a performer who climbs a ladder of swords barefoot; an act in which "Madame Electra" is strapped to an electric chair, which is then charged; and a "blade box," in which a female performer lies in a wooden box whose sides are progressively intersected by blades.

Fig. 15: Coney Island Side Show banner-line (2005). Courtesy Michael Bolger.

As this list indicates, the majority of Coney Island's current contemporary performers are, in keeping with the shift noted at the start of this chapter, "self-made" freaks rather than "born" freaks. However, Dick Zigun has continued to actively recruit performers with unusual forms of embodiment, employing Otis Jordan after he left the Sutton Side Show; Jennifer Miller, a woman with a beard who performed "Zenobia, the Bearded Lady," and who has described being "lured" into the sideshow by Zigun (Carr 59); and Mat Fraser, the well-known British actor and playwright with phocomelia (or foreshortened arms), who once took to the Coney Island stage with the word "freak" penned on his chest, and has continued to participate in the sideshow's Super Freak weekends. The most controversial of Coney Island's recent acts, however, has been that of Koko the Killer Clown, a "rotund, developmentally disabled dwarf," as Rachel Adams describes him, who, Michael Chemers notes, "regards the audience with silent malice" (2). Public reactions to Koko's performances are, notes Adams, largely negative:

> Koko was the only performer I saw at Coney Island who might have earned a job as a freak in earlier days.[...] Lumbering onstage with his face slathered in black and white grease paint, Koko delivered a loud,

atonal monologue as he struggled to make balloon animals. My response, shared by other audience members, was distinct discomfort.[...] The source of our revulsion must be the way that Koko, unlike other performers, recalls the aspects of the past freak shows that would be most intolerable to a contemporary viewer. Some line between individual agency and exploitation had been crossed, and the spectator who has paid for the pleasure of looking becomes an unwitting accomplice. (216)

Although academic researchers do need to be careful not to homogenise audiences' responses or to attribute their own reactions to those of the diverse public around them, Adams' assertion that audiences were made uneasy by Koko's act and that his evident developmental disability, rather than his unusual anatomy, was likely to be the main cause of this unease, is probably accurate. Despite the evident discomfort his performances caused, however, Koko has continued to perform at the Coney Island Side Show, appearing during the annual "Congress of Curious Peoples" in 2009, 2010 and 2011. Moreover, while questions about Koko's ability to give informed consent to his performance are certainly important, and recall the issues raised above by David Gerber in his critique of Otis Jordan's participation in the Sutton Side Show, the discomfort of contemporary audiences with developmentally disabled or congenitally unusual bodies in public or performance spaces does not necessarily attest to improved sensitivity or social conditions for these subjects on or off the stage. Rather, it may just represent an unwillingness by mainstream publics to see such bodies or to be confronted by the difficult issues (about ethics, about access, about the limits of choice and self-determination) their existence poses, especially in sites, like Coney Island, dedicated to relaxation and recreation.

Many of the questions raised by Koko's act are directly addressed in the work of Mat Fraser, which constitutes a sustained and explicit engagement with the history of the freak show. Fraser has compiled a significant body of work that reflects critically on the history of the freak show while providing a thoughtful commentary on the tensions between the concepts of freakery and disability that underpin this. Fraser wrote and starred in *Thalidomide: The Musical!* and *Sealboy: Freak*, which recreated the life and sideshow act of the 1950s performer Sealo, Stanley Berent [Fig. 16]. In 2002 he also filmed a documentary called *Born Freak* for Channel 4 in the UK. His current projects include *The Freak and the Showgirl* (a collaboration with the neo-burlesque performer Julie Atlas Muz). While Fraser's engagement with the freak show does not raise the questions about agency posed by Koko's participation in Sideshows by the Seashore, criticism of Fraser's work reproduces some of the same assumptions that underlie critical discomfort with

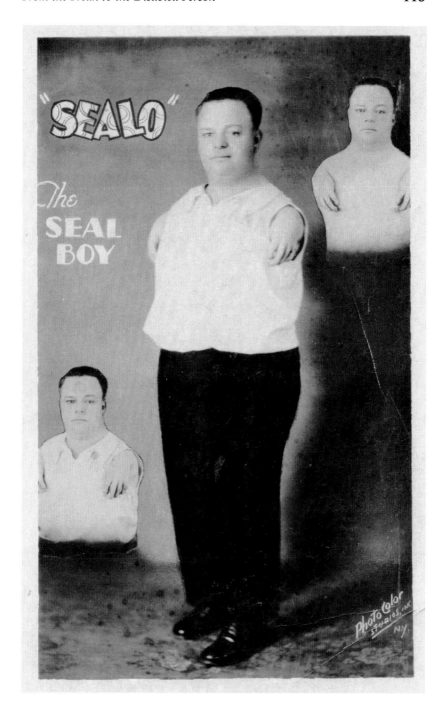

Fig. 16: Sealo the Seal Boy promotional poster (undated). Courtesy of James G. Mundie.

Koko's performance. For Mitchell and Snyder, Fraser's performance at Coney Island (filmed as part of *Born Freak*) represents less a critical engagement with the freak show than an exposure to an institutional and discursive system of spectacularisation that he cannot control, and which ultimately defines his body in negative ways that he does not have the power to reinscribe. On the Coney Island stage, Mitchell and Snyder claim, Fraser

> experiences his own inevitable degradation. His stage performance, which literally recreates an exhibition of one of his nineteenth-century freak show ancestors, Seal-o, fails to achieve the desired level of political satire. In the process of duplicating the comments and actions of his predecessor, we watch as the act increasingly mires the performer in degrading spectacle. Fraser finds no effective politicized venue within the carnival tent and, as if to emphasize this fact, the camera performs various pans across the faces of Fraser's audiences; they stare at the performance with a collective discomfort and the show seems almost too humiliating to witness from an "enlightened" freak show audience's perspective. The act of simply occupying an objectifying gaze is no longer possible—if it ever was to begin with—and the performer's audiences are caught either looking away in embarrassment or staring with some difficulty. (http://www.dsq-sds.org/article/view/575/752)

While Mitchell and Snyder's conclusion about Fraser's experiences at Coney Island is somewhat undermined by the fact that he has continued to perform in this space—most recently alongside Koko the Killer Clown at the 2011 Congress of Curious Peoples—their criticisms are also a clear indication of how problematic the freak show of the twenty-first century continues to be, and how fraught are the issues it raises.

Despite the difficulties and discomfort it often causes, contemporary performers with unusual forms of embodiment continue to be attracted to the freak show stage, and to Coney Island in particular. The reasons for this are directly addressed by Jennifer Miller in her account of her experiences performing as a "bearded lady" at Sideshows by the Seashore [Fig. 17]. When Miller, now director and MC of New York's queer performance group Circus Amok, was interviewed about her time as part of the Circus Sideshow, she explained that her decision to accept this role was motivated by a desire to engage directly with the cultural figure through which her own physical difference is often understood:

> I was growing my beard long before I worked in the sideshow, so I always had this image of the bearded lady as kind of this little icon sitting on my shoulder, you know, battling with me and how I was seen in the world. So when the opportunity came, when I was invited, enticed, to come work in the sideshow, I wanted to give it a try. I wanted to meet

Fig. 17: Jennifer Miller in front of her Coney Island "Sideshows by the Seashore" banner (1998). Courtesy of Tami Gold Coney Island Side Show, NYC.

this person, this image, this history that I had been in dialogue with, sort of face to face. (*"Freaks": The Sideshow Cinema*. Warner Brothers, 2004)

For Miller, then, Coney Island provided a space in which to meet, and confront, the figure of the "bearded lady" through which her body, and those of other women with beards, are still understood. In choosing to step into this role, to literally embody it, Miller was on the one hand making the same compromise that had been made by generations of people with unusual anatomies before her. On the other hand, however, her performance attests to the institutional changes the freak show has undergone since its revival in the early 1980s. Whereas earlier incarnations of Coney Island's sideshow, such as Dreamland and Wonderland, used inside and outside talkers to provide descriptions of their attractions and their acts, during Miller's performance she spoke for herself in a way that allowed her to talk back, critically, to the tradition that had constructed public perceptions about her corporeal difference.

This change—the move from using talkers to allowing performers to speak for themselves—represents an important and epochal shift in the nature of the freak show, representing a profound transformation from its nineteenth- to twenty-first century formats, in a way that enables twenty-first century performers to critically interrogate the traditional concepts of bodily difference by which the cultural significance of their own bodies are still largely determined. Katy Dierlam, who performed at Coney Island as the "fat lady" Helen Melon, narrated as part of her act the history of Dolly Dimples, "the most famous Fat Lady" of the sideshow, who performed at Coney Island in the 1930s (Mazer 268). Eak the Geek, the inside talker for Sideshows by the Seashore until 2006, closed the show by cautioning the viewer that the figure of the freak exists only on the sideshow stage, as a result of the audience's spectatorship and construction of the performer's body. "In the world outside," he reminded his viewers, "there are no freaks." Thus, while performing as the "bearded lady" or "fat lady" can be seen as participating in the ongoing circulation of often oppressive cultural figures of bodily difference, stepping into these roles can also provide the members of the circus sideshow with the opportunity to talk back to those figures, challenging their cultural dominance.

If so many contemporary performers have chosen to participate in self-identified freak shows like Coney Island, it is because they recognise the ongoing importance of the sideshow stage as a space in which popular ideas about their own embodiment have historically been constructed. As Miller's account of her own relationship to the figure of the bearded lady makes clear, whether or not she chooses to perform this role on the sideshow stage, this is still the figure through which her own bodily difference is popularly understood. Moreover, on stage, her audience recognises that they are witnessing a performance; on the street, Miller is still likely to be seen by passing strangers as a bearded lady, without any awareness of the extent to which this is a constructed, theatrical category. The sideshow stage thus affords opportunities to talk back to and critique this figure that do not readily exist outside it. In this respect, Miller's performance as Zenobia, which exists in tension and critical dialogue with the figure of the bearded lady, is exemplary of the way performers in twenty-first century freak shows critically reflect on the tradition they also continue, interrogating the cultural assumptions about embodiment that continue to circulate in and through this space. The ongoing appeal of the sideshow for performers and audiences alike thus resides in the potential that Miller and others have recognised and exploited in their acts: sideshows are sites in which norms about the body, its limits and capabilities, are theatricalised and transformed into spectacle, but in which, for this very reason, they can also be challenged and contested.

Despite its long and recognised history as a semantically and culturally volatile space, one in which the meanings of anatomically unusual bodies are often debated and contested, as we have seen in this chapter, much of the discussion about the significance of the freak show has attempted to stabilise its effects, to resolve the question of what kinds of ideas about bodies and bodily difference the sideshow stage ultimately (re)produces. In Petra Kuppers' account, the freak show is not a site of destabilisation but, on the contrary, "a spectacle of certainty: even if in other areas of life uncertainty rules, here, one's position on the far side of the stage is assured" (*Disability and Contemporary Performance* 34–35). Thus, while "the psychic effects of the freak spectacle have destabilising effects, assaulting the boundaries of firm knowledge about self," this element of the freak show serves only to better "strengthen them again in cathartic effect" (45). Kuppers thus reads the freak show in exactly the same way that Bakhtin has famously read the carnival—as a space that seems to operate outside the bounds of conventional cultural expectations but which ultimately functions to better reinforce them. For Bakhtin, the (pre-modern) carnival was a space in which the values and practices of the everyday world were suspended and overturned; it was a "world upside-down" that was sanctioned by the dominant order as an authorised "safety valve," because it served ultimately to assure its continuity and power.

The nineteenth- and twentieth-century sideshow is often understood in almost identical terms. Thus, while John Kasson argues that Coney Island provided nineteenth-century New Yorkers with "an area in which visitors were temporarily freed from normative demands ... a relatively 'loose,' unregulated social situation which ... broke down the sense of rigidity that dominated so much of the life of American cities at the turn of the century and lessened personal restraints" (41), Jon Sterngass contends that this served merely to regulate and commodify practices of pleasure in a way that reinforced the status quo: "The liminal world is by definition a temporary moment in time, and despite the emphasis on breach and collectivity, the resort journey concluded with the reintegration of travellers into an unchanged American society" (272). For this reason, Sterngass argues, a journey to Coney Island "fortified rather than subverted the strictures of everyday life. Instead of idiosyncratic laboratories for the rethinking and reworking of American norms, resorts had been transformed into mundane places where visitors bought souvenirs, gambled for money, partied in private mansions" (277).

Despite Sterngass's claim that the carnivalesque space of the nineteenth- and twentieth-century funfair ended by reinforcing the economics and cultural dynamics of bourgeois life, as we have seen above, the sideshow stage was a space in which traditional power dynamics were acted out in ways that explicitly challenged these or problematised their dominance. We can see this by returning to the sideshow proprietor with whom we began this chapter, P. T. Barnum. One of Barnum's most influential—and well-recognised—contributions to Victorian popular culture was his inventive use of mass-marketing, in which he mischievously advertised his exhibits as fraudulent and played up his own predilection for "humbug." Barnum's exhibition of Joice Heth exemplifies this. On his first tour with Heth to Boston, Barnum discovered he had rented an exhibition room next door to one of the most popular and celebrated curiosities of the day—Johann Maelzel's automaton chess player. Shortly afterwards, "anonymous" notices began to appear in the local press, questioning whether Heth really was the prodigiously old woman she was advertised to be, or whether she was actually an automaton herself. Audiences thronged to the exhibition, many arguing that Heth was quite obviously a mechanical figure. "What made Barnum's new (and seemingly counter-productive) marketing scheme innovative," argues James Cook, was his recognition that

> artful deception was never a hard and fast choice between complete detection and total bewilderment, honest promotion and shifty mis-representation, innocent amusement and social transgression. Rather, Barnum suggests, it was precisely the blurring of these aesthetic and moral categories that defined his brand of cultural fraud and generated much of its remarkable power to excite curiosity. (16)

This atmosphere of trickery and hoaxing constituted the freak show as a space in which the usual power relations between the (dominant) spectators and (marginal) freaks could be reversed. In this way, as we have seen in the earlier sections of this chapter, the sideshow stage was a volatile space in which meaning was, and continues to be, constantly (trans)formed and reconfigured. Thus, while the nineteenth-century freak show certainly earned its reputation as a space in which the poor and vulnerable with unusual anatomies were exploited for profit and for the amusement of a leering crowd, they were also sites of the reciprocal exploitation of the audience by the performers, particularly through fake exhibits, or gaffs, that served to destabilise the category of the "freak" even as they constructed it. (Half-men/half-women performers, for example, were almost always fakes.) It is well known that the eleventh exhibit in a traditional ten-in-one sideshow is called the "blow-off," an obviously gaffed exhibit designed to remove one audience from its seats and to make way for new paying customers. Barnum's crowded American Museum, for instance, had a sign that read "This way to the Egress," and led his surprised visitors, who were expecting an additional exhibit, back out onto the street.

It is precisely this aspect of the freak show—its capacity to conspicuously trick or wrong-foot its audiences, to explicitly announce that it is gulling its audiences even as it successfully does so—that makes it such a productive space for contemporary performers, and it is this dynamic to which their own critical interrogations of traditional figures of freakery often call attention. If contemporary performers like Fraser and Miller have found in the freak show a productive space within which to reflect on the historical and cultural conditions in and through which perceptions about their own physical difference have been constructed, it is primarily for this reason: because the freak show itself, even in its most traditional form, contains the potential for different kinds of signification and different dynamics of power. This potential has only been intensified by the much-heralded "death" of the traditional freak show in the mid-twentieth century, which has served to evacuate this site as an ongoing tradition and thus make it available for appropriation and resignification. Rather than answers about bodily difference or representing the increasing dominance of professional medical discourses over popular spectacular ones, then, what we find on the freak show stage is the ongoing reconfiguration of the scientific and spectacular through the changing exhibitory culture of the nineteenth and twentieth centuries.

Notes

1 As Chemers notes, although Barnum himself never used the word "freak," his exhibitions of anatomically unusual bodies as a series of types—the bearded lady, the dwarf or giant, the pinhead, the albino, the nondescript, etc.—would come to define the "freak" as a new category of public exhibit and the "freak show" as a new kind of spectacular space (1–9).

2 As Tony Bennett notes: "the founding collections of many of today's major metropolitan museums were bequeathed by international exhibitions; techniques of crowd control developed in exhibitions influenced the design and

layout of amusement parks; and nineteenth-century natural history museums throughout Europe and North America owed many of their specimens to the network of animal collecting agencies through which P.T. Barnum provided live species for his various circuses, menageries and dime museums" (5).

3 Pastrana, whose medical condition has since been identified as "autosomal dominant syndrome of congenital hypertrichosis and gingival hyperplasia" (Bondeson 243), was so popular a performer that her mummified remains continued to be displayed long after her death, right through to the 1970s, when they were vandalised and subsequently moved into storage at an anatomical institute in Oslo.

4 Many of Barnum's exhibitions, like those found in contemporaneous World's Fairs, were of exoticised bodies: "wild men" and "pigmies," Chinese families and Laplanders, Circassian beauties and the "Hottentot Venus". While obviously a related practice, the exhibitory styles and interpretative frameworks according to which these bodies were made the objects of public display were significantly different from those with which this chapter is concerned. Histories of the exhibition of exoticised bodies in nineteenth-century circuses and sideshows can be found in Roslyn Poignant's *Professional Savages: Captive Lives and Western Spectacle*, Jane Goodall's "Acting Savage," and Barbara Kirshenblatt-Gimblett's *Destination Culture: Tourism, Museums, and Heritage*.

5 This practice appears to have been first introduced in the late eighteenth century. Flyers for the exhibition of "human curiosities" in the popular exhibition halls from this time begin to include a list of the "members of the medical profession" alleged to have attended the exhibition and inspected the performers' bodies.

6 Teratology developed as an increasingly important field of research in the mid- to late nineteenth century, catalysed by the publication of Isidore Geoffroy de Saint-Hilaire's *Treatise on Teratology* (1832).

7 The body of Jeremy Bentham (the social reformer who developed the idea of the panopticon) was dissected as part of a public anatomy lecture, then preserved and put on public display. Bentham's body can still be seen at University College London, which acquired his remains in 1850.

8 Leslie Fiedler's *Freaks: Myths and Images of the Secret Self* (1978) predates Bogdan's study, but specifically contextualises the freak figure within the countercultural movement of the 1960s and 1970s.

9 A ten-in-one sideshow, as its name suggests, is a sideshow in which a single ticket provides entry to a performance consisting of ten separate acts. When performed in continuous rotation, as at Coney Island, this is known as a "grind show." The performers stage up to 10 shows a day, and 700 over the season's run. While circuses and freak shows are not the same thing, each of the circuses listed here draws on the style and acts found in nineteenth-century freak shows, and many of their performers have prior experience in freak shows.

10 Descriptions of "pygmies" and "wild men" exhibited in the mid- to late nineteenth century suggest that a number of these performers were suffering from conditions such as congenital hypothyroidism, or "cretinism," caused by a lack of iodine in the diet. As Armand Marie Leroi notes, "the legislated spread of iodized salt in the early twentieth century eliminated European goiter and cretinism within a generation, so that today these diseases are little more than folk-memories" (198).

11 It should be recognised that the changing economics of the sideshow have also played an important role in this respect. Nickell argues that the increasing popularity of mechanical rides in the mid-twentieth century also contributed

to the end of the freak show: "Hall states that the decline of sideshows began in the mid-1950s with the advent of the big rides," writes Nickell. "He says, 'They were like a vacuum cleaner. They'd just suck money up off the midway'" (Nickell, 346; see also Siegel 107). As Dick D. Zigun noted in *The New York Times*, the subsidised reopening of the Coney Island Side Show was motivated in part by a desire to preserve Coney Island itself from a similar, perhaps terminal, decline: "Coney Island was at the crossroads," Zigun explained. "It was getting dangerously small and could have been rezoned for residential housing." (quoted in Lee 6).

12 However, a promotional film made in this very year, *Coney Island 1940*, shows the Dreamland sideshow thronged with people, and an outside talker touting for business by placing the show's resident "pinhead" (or microcephalic) on the bally platform.

13 David Williams' *Deformed Discourse: The Function of the Monster in Mediaeval Thought and Literature* provides a detailed account of the central role played by ideas about monstrosity in the epistemologies, religious discourses and literary texts from the late medieval period into the early modern period.

14 Ambroise Paré (c. 1510–1590) was the royal surgeon under four successive kings in sixteenth-century France, and is still widely recognised as the founder of modern surgery. He was also a pioneering anatomist who invented several surgical instruments and experimented with early prostheses.

15 Maternal impression is the belief that women's actions, sensations and thoughts during pregnancy played a formative role in shaping their unborn children: a bad shock, an unexpected sight or uncontrolled emotions were among the main attributed causes of monstrous births for early modern writers.

16 Bulwer (1606–1656) completed four other studies, of which the best known is *Chirologia: or the naturall language of the hand. Composed of the speaking motions, and discoursing gestures thereof. Whereunto is added Chironomia: or, the art of manuall rhetoricke. Consisting of the naturall expressions, digested by art in the hand, as the chiefest instrument of eloquence* (1644). He is believed to have been awarded a doctorate in medicine shortly before his death.

17 For accounts of the politicisation—as opposed to medicalisation—of the monster in early modern thought, see Laura Lunger Knoppers and Joan B. Landes' edited collection *Monstrous Bodies/Political Monstrosities in Early Modern Europe*, Jeffrey Cohen's collection *Monster Theory: Reading Culture*, and Peter Platt's edited volume, *Wonders, Marvels, and Monsters in Early Modern Culture*.

18 Michael Chemers' *Staging Stigma: A Critical Examination of the American Freak Show* provides a much more detailed account and critique of this history, noting, importantly, that the later "1903 version of the revolt was almost certainly a publicity stunt, a hoax staged to capture the imagination of turn-of-the-century London and to deflect would-be censors. In the bait-and-switch world of freak shows promotion, where nothing can be taken for granted, this is probably very likely" (100).

19 Jordan's act included using his mouth to roll, light and smoke a cigarette, part of an established centuries-old tradition of such performers, including Madame Rosina, who was born without arms but could crochet with her feet and paint works of art with her mouth, or the prominent eighteenth-century performer Thomas Inglefield, who "by industry acquired the Arts of Writing and Drawing, holding his Pencil between the Stump of his Left Arm and his Cheek and guiding it with the Muscles of his Mouth" (*A Curious Collection of Prodigies* 5).

20 While it is increasingly recognised that vernacular photography of this kind provides an important historical record of popular culture in the nineteenth

century, as Darrah and Joel-Peter Witkin have recognised, it was for a long time disparaged in and by histories of nineteenth-century photography and has only recently been reappraised for the important role they played in Victorian popular culture. See, for instance, Allan Sekula's "The Body and the Archive" or Abigail Solomon-Godeau's *Photography At The Dock: Essays on Photographic History, Institutions, and Practices*.

21 See Darrah 136 or Dennet 114. Mathew Brady, one of the most prominent American photographers of the nineteenth century, although famous for his Civil War images, also took many photographs of freak performers.

22 Mills, born in 1860, was probably on tour in sideshows by the late 1870s. She retired in 1890 due to ill health and died the following year.

23 Garland Thomson refers to this process as "enfreakment": the transformation of physical difference into something recognisable as that of a freak (*Extraordinary Bodies*). As Bogdan explains: "being extremely tall is a matter of physiology—being a giant involves something more [...] the enactment of a tradition, the performance of a stylised presentation" (3).

24 Mills suffered from Milroy disease, which restricts development of the lymph vessels in the legs and causes fluid build-up.

25 Landouzy's *Traité complet de l'hystérie* (1846), Brachet's *Traité de l'hystérie* (1847) and Briquet's *Traité clinique et thérapeutique de l'hystérie* (1859) all preceded, and enabled, Charcot's own writing on hysteria. (See Edelman 7–14, and Beizer 30–54 for accounts of Charcot's indebtedness and contribution to the wider nineteenth-century conceptualisation of hysteria.)

26 Didi-Huberman also notes that Charcot was an avid collector of anatomical museum catalogues (30).

27 Allan Sekula examines the complexities of this dual understanding of the archive in "The Body and the Archive" (3–64).

28 Lord's own reading of this Foucauldian account of the museum is critical. Drawing on Foucault's examination of the museum in "Of Other Spaces: Heterotopias," Lord argues that the heterotopic character of museums means that these cannot be identified as essentially disciplinary or resistant, progressive or conservative, but, rather, always incorporate both these possibilities.

29 Bosker and Hammer note that this style of banner art, while widely associated with the freak show, dates only from the 1930s.

CHAPTER 4

Inventing the Bodily Interior: Écorché Figures in Early Modern Anatomy and von Hagens' *Body Worlds*

When Amanda Wilson staged a naked demonstration at the 2002 *Body Worlds* exhibition in London, as a protest against the show's gendering of standard anatomy as male, the conventions governing the modelling and public display of human anatomy were over three centuries old. As we have seen in the previous chapters, conventionally, male figures are used to represent "standard" or "normal" anatomy, whereas female figures show only the female reproductive system and foetal development. The continuation of this tradition in *Body Worlds* attests to its enduring nature and its ongoing influence on contemporary anatomical modelling. We see this in Gunther von Hagens' claim that the reason *Body Worlds* uses male plastinates to represent every aspect of anatomy except reproduction is because women's bodies, with their higher percentages of body fat, do not show the musculature of "the body" clearly enough. However, Wilson's actions did provoke a direct response from von Hagens, who subsequently began to produce female plastinates in active poses, as can be seen in newer exhibits such as "The Archer" (2005). While this model clearly represents an attempt to include more gender-neutral figures for standard anatomy, thereby breaking with the long tradition in which female anatomy is confined to its reproductive system, the modelling of "The Archer" also demonstrates how resilient these conventions are. Although this model, like all of von Hagens' whole-body plastinates, is an écorché figure (that is, one from which the skin has been removed), and although it is designed to show the muscles of the arms and shoulders in action, the fat and flesh of the breasts of "The Archer" have been retained and the nipples reattached, thereby emphasising its gender by feminising its form. Moreover, as gender remains one of the key cultural frameworks through which we are trained to view the body and to understand its cultural significance, anatomical models cannot simply be de-gendered; rather, assumptions about gender will continue to inform—in more or less obvious ways—the aesthetics of anatomical modelling. The extent to which the anatomical vision that has shaped the modelling of "The Archer" remains itself shaped by the same conventions that produced the anatomical Venuses of the eighteenth century can here be clearly seen.

At the same time, in locating her protest within *Body Worlds* itself, Wilson's actions call attention not only to the conventionality of the anatomical vision of the body, which we have been examining throughout this book, but also to the equally long history in which such exhibitory spaces have provided sites in which the dominant ways of seeing and knowing the body have been critically examined and contested.[1] As we have seen in the previous chapters, the ideas about bodies (re)produced in and by popular anatomical exhibitions are not fixed and stable, but are rather temporary formations that arise within the context of much wider and equally dynamic institutional and discursive networks. Like these earlier exhibitions, the bodies on display in *Body Worlds* have a tendency to problematise the very thing they claim to represent.

We have already seen this in detail with relation to gender in chapters 1 and 2, which examined, respectively, the way that anatomical models encode changing ideas about femininity and masculinity, with particular kinds of models becoming popular objects of public display during periods in which cultural assumptions about those bodies came under heightened scrutiny. This chapter, accordingly, will not focus on the gender of anatomical modelling, but rather on a kind of body, or a way of seeing bodies, that remains much more naturalised within the space of the *Body Worlds* exhibition and yet is much more problematic: the "real human bodies" that von Hagens repeatedly claims to put on display in an unmediated way. Interrogating von Hagens' central assertion that his plastinated figures are not "models," which would be "nothing more than an interpretation" (www.bodyworlds.com), but instead depict anatomy itself, this chapter will examine the significance of von Hagens' own contextualisation of his work as a continuation of the tradition of anatomical art that starts in the early modern period. This is the moment when the écorché figure becomes central to anatomical illustration. Examining the precedence of the early modern écorché figure in shaping not only the iconography but also the anatomical knowledge represented by von Hagens' plastinates, this chapter will position von Hagens as the heir to a long anatomical tradition for which the removal of the body's skin and the exposure of its interior is understood to reveal the "truth" of its condition and significance. As such, von Hagens' work in many ways represents the apotheosis of the anatomical vision whose emergence and subsequent effects we have been examining throughout this book. However, von Hagens' plastinates do not simply reveal the inside of the body, as we will see; rather, they are exemplary of the invention of a specifically modern concept of bodily interiority, which is central to the construction of the individuated and self-managing subject emergent at the same time.

Anatomy as Spectacle

Among the many *Body Worlds* exhibitions staged in 2009 — in Brussels, Cologne, Zurich, Manchester, Houston, Philadelphia, Toronto and Singapore — the Manchester exhibition is particularly notable in that it marks the first occasion on which von Hagens has returned to London since 2002. This was the year von Hagens staged his notorious public autopsy in an East End art gallery, which was subsequently televised by Channel 4.[2] Von Hagens' event was the first time a public dissection had been held in Great Britain since the establishment of the Anatomy Act in 1832, and was undertaken despite concerns by the Queen's Inspector of Anatomy, Dr Jeremy Metters, that the autopsy would be illegal under the Anatomy Act.[3] Like the nineteenth-century public autopsy of Joice Heth, staged by Barnum, which was discussed in the previous chapter, this one was performed in front of a mixed audience. In Helen MacDonald's account, the medical professionals present were left largely disappointed by the proceedings: "They noticed when he struggled to saw open the skull, and seemed unable to locate the pancreas" (7). The general public, on the other hand, were highly entertained: "When von Hagens pulled out the sternum with both hands and plunged into [the cadaver's] thorax to lift out his heart and lungs, people cheered" (7).

Rather than focussing on the issue of whether von Hagens is really a competent anatomist or merely a showman, this chapter begins by considering von Hagens' early framing of his practice as "event anatomy" or "anatomy art." In the early 2000s, von Hagens repeatedly worked to blur the boundaries between the medical and spectacular elements of his show, rather than attempting to clarify them. His predilection for intensifying the public debate about his work rather than attempting to resolve it is evident in his 2002 interview with the *Guardian* (referenced in the introduction to this book) in which he openly acknowledged inciting provocation for strategic and commercial purposes:

> von Hagens giggles. "It is an honour to cause this controversy," he says as he strides through the Brick Lane puddles.[...] [H]e is part shaman and part showman; at once an anatomical scientist bent on shaking up a western society that he regards as living in denial of its corporeality and of death, and a PT Barnum basking in the media hoopla of his British reception, aware that part of the appeal of Body Worlds is the same as that which drew our ancestors to public executions and freak shows.[...] Despite all this, he revels in the fuss. "I don't mind if you're sensationalist in your article," he says. "More people will come if you are." (Jeffries 3)

This element of staged controversy can be found throughout von Hagens' work. His television series, *Anatomy for Beginners*, made for Channel 4 in the UK, includes close-ups of the audience grimacing and recoiling while dissections are carried out on stage. In the documentary *The Anatomist*, interviews with von Hagens, cheerfully identifying himself as "The Walt Disney of Death," are interspersed with those of censorious British medical figures explaining why they strongly disapprove of his work. Given the control von Hagens exerts over the publicity and reproduction of his work, and the fact *The Anatomist* is included as part of the DVD of *Anatomy for Beginners*, such criticisms must be understood not as independent or external to von Hagens' promotional rhetoric but rather as part of *Body Worlds'* own packaging of itself as controversial. Similarly, while exposure to the actualities of an autopsy is undoubtedly confronting for many spectators, the inclusion of the shocked reactions of the audience in *Anatomy for Beginners*—that is, within the context of an edited television show—constitutes a deliberate construction of such events as both confronting and titillating.

However, since its relocation to the United States in 2004, *Body Worlds* has undergone a dramatic cultural repositioning. While his exhibitions were originally held in commercial exhibition halls or convention centres (like the Brick Lane art gallery he used in London's East End), in the USA, the show has exhibited exclusively within science museums, and von Hagens has increasingly identified himself as a health educator whose central purpose is to "democratise" anatomy.[4] His advertising material now utilises the rhetoric we examined in the introduction to this book, stressing that the aim of the exhibition is "to educate the public about the inner workings of the human body and show the effects of poor health, good health and lifestyle choices.

They are also meant to create interest in and increase knowledge of anatomy and physiology among the public" (www.bodyworlds.com). For Linda Schulte-Sasse, this increasing emphasis on the educative and scientific purpose of the exhibition, coupled with a corresponding de-emphasising of its commercial, spectacular elements, has lent the North American version of the exhibition "a prestige and respect-ability it lacked in Europe, an institutional endorsement as serious science" (372).[5] This is achieved, to a large degree, by re-establishing the binary between serious medicine and the sensationalist spec-tacle that von Hagens' earlier publicity so gleefully set out to undermine—that is, by insisting that *Body Worlds* is a serious lesson in human anatomy *as opposed to* a spectacle, that it provides a potential for learning about the body rather than an occasion to gawp at it. As a result, in the USA, "the show is cloaked in a mantle of solemnity, respectability and high culture, assuring that *Body Worlds* is a celebra-tion of enlightenment, scientific progress and altruism" (Schulte-Sasse 374). While von Hagens' work remains controversial, then, as we saw in the introduction, the large number of exhibitions he continues to stage serves as testimony to his success in culturally repositioning his work.[6] As such, his exhibitions trace the reverse trajectory to the one we have seen throughout this book, which defines the history of all other anatomical exhibitions. That is, whereas earlier exhibitions were launched with claims that they instruct and edify their audiences, claims often widely supported in the press, only to later slide into disrepute, von Hagens' exhibitions began in a swirl of Barnumesque controversy and negative publicity, only to become more accepted and respectable over the years, metamorphosing "from an enticing but dubious form of entertainment into the ideal family-friendly Sunday outing or junior-high field trip" (Schulte-Sasse 371).

Von Hagens has largely achieved this by contextualising his work culturally, as reinforcing contemporary discourses about public health education, and also historically—positioning his work within the long tradition of anatomical art stretching back to the Renaissance, thereby lending it an additional authority and legitimacy. The plastinates in his exhibitions are surrounded by screen prints of early modern anatomical art, situating these within the long tradition of representing écorché figures in anatomical study and thus providing a historicised context within which spectators can interpret the cultural as well as anatomical significance of the plastinates on display. Uplifting quotations about the body and mortality from religious and philosophical sources (including the Bible, Shakespeare, Kant, Goethe and Nietzsche) are also printed on large posters hung about the room.

One of the most striking examples of *Body Worlds'* indebtedness to, and citation of, early modern anatomical art is the positioning of the "Skin Man" plastinate (1997) alongside a screen print reproduction of the iconic Valverde illustration, "Muscle Man" (1556) [Fig. 18]. Juxtaposing these two figures in this way is clearly designed to present the plastinate as the realisation and embodiment of the anatomical illustration it replicates, providing a "real" three-dimensional model that supersedes Valverde's drawn image, and thus triumphantly materialising what, in earlier centuries, could only be depicted with pen and ink. At the same time, however, von Hagens himself claims that his work is of a fundamentally different order to these earlier forms of anatomical art. Whereas earlier anatomical models or illustrations were artistic *representations* of the body, von Hagens argues, his plastinates present "real human bodies," human anatomy itself, in an unmediated way: "[n]either illustrations nor models can convey the individual beauty of these structures to us, for the source of truth is in the originals" (www.bodyworlds.com). Reflecting on this aspect of his work, José Van Dijck argues: "von Hagens hails his technology as enabling direct inscription, eradicating all mediation between object and representation [...] thus perpetuating the myth of 'a scientific transparent truth' — a pure representation of the human body without the contamination of human intervention" (117). The fact that his exhibition features "real human bodies" thus makes *Body Worlds*, according to its own publicity, not only the realisation of, but also superior to, the tradition of anatomical art by which it is preceded.

Despite his claims to represent "real bodies," von Hagens' use of an early modern illustration of an écorché figure as model, and the fact that this early modern tradition has very clearly shaped the aesthetics of his own work, serves the contradictory purpose of revealing how conventional these representations of "real bodies" are, how mediated by traditions of seeing that body developed over many hundreds of years within anatomical modelling. In particular, two key conventions identifiable with early modern anatomical art are reproduced in and by von Hagens' plastinates. The first of these, evident in the modelling of "Skin Man," is the practice of representing the anatomical subject with open eyes, alive and conscious of the dissection being carried out on his/her body. Petherbridge and Jordanova identify this as one of early modern anatomical art's most persistent visual tropes (27). As Jonathan Sawday notes, "the conventions of anatomical illustration demanded that the figure, even at the very deepest stages of dissection, should be represented as still alive" until well into the eighteenth century (112). The second striking convention von Hagens inherits from early modern

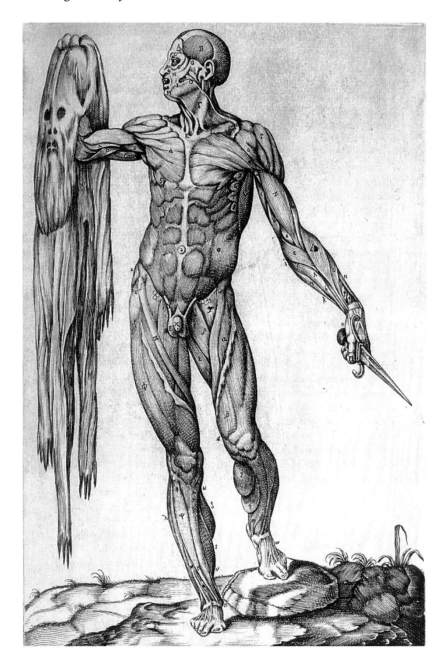

Fig. 18: Valverde's "Muscle-Man". In *Historia de la composicion del cuerpo homano* (1556). Courtesy of the Wellcome Library.

anatomical art is the representation of écorché figures as active participants in their own dissection, holding the knives by which they were
flayed, or holding open their own flesh to expose the interior anatomy.
Valverde's "Muscle Man," for instance, is depicted holding the knife by
which his skin has been removed. Jonathan Sawday argues that such
images represented a body docilely complying with its own dissection,
willingly acknowledging itself as that passive subject of anatomy, much
as did the Venuses examined in Chapter 1: "the illustrations show us
a corpse conspiring with its own demonstration, in order to confess
the truth of the study which has been embarked upon. The body, in
other words, complied in its own reduction and thereby confessed to
the power and truth of this new branch of human enquiry" (113–14).
This representation of a docile subject of dissection, who is a willing
participant in the acquisition of anatomical knowledge, is reproduced
in von Hagens' work, not only in the modelling of his plastinates but
also in the publicly displayed notices emphasising that all the bodies on
display in his exhibitions have been voluntarily donated. Information
about body donation features prominently at all of his shows.[7] In the
entrance way to the 2006 Philadelphia exhibition, a large sign informed
visitors: "The specimens in this exhibit are Body Donors, individuals who during their lifetime bequeathed that upon their death their
bodies could be used in this exhibition.[...] For their clear vision and
tremendous generosity we are deeply thankful." (It should be recognised that this emphasis on the voluntary nature of the body donation
process within the space of the exhibition is clearly motivated by the
need to address ongoing concerns about Von Hagens' ethics and the
source of the bodies on display).[8]

While von Hagens explicitly draws on the traditions of early modern
anatomical art, and represents himself as continuing its tradition of
public exhibitions of anatomy, in contextualising his work in this way
he also, and paradoxically, serves to undermine his central claim that his
plastinates demonstrate anatomy "itself," in a manner "free of interpretation." The constructedness and conventionality of both traditional
écorchés and von Hagens' plastinates attest to the fact that neither of
these forms simply represent the anatomical "truth" of the body in an
unmediated fashion: these images are not an objective record of the
current state of anatomical knowledge in any given period: rather, they
encode historically specific assumptions about the nature of the body
and of knowledge itself. We can consider this further by examining the
way this early modern tradition was invoked by von Hagens during
the staging of the public autopsy he performed in London. Before the
dissection began, von Hagens again contextualised his practice within

the long history of public anatomy, explaining to the audience "that he was merely following in the tradition of the great Renaissance anatomist Andreas Vesalius, who educated the world with such procedures in the 16th century" (Petropoulos 142). This assertion was reinforced by the prominent display of Rembrandt's 1632 painting *The Anatomy Lesson of Dr Nicholas Tulp*, hung at the back of the dissection theatre and thus presiding over the autopsy. The black hat worn by von Hagens, matching that worn by Dr Tulp on the canvas, further emphasised the continuities between Rembrandt's image and von Hagens' act. Despite this careful framing of the dissection, however, descriptions of this event suggest that, rather than resembling the (private, professional) dissection depicted by Rembrandt, von Hagens' autopsy instead evoked another famous anatomical scene, that of the frontispiece of Vesalius's *De Humani Corporis Fabrica* (1543). In contrast to Rembrandt's representation of the anatomy theatre as an intensely professionalised space, in which the doctors gaze with sombre concentration at the cadaver to be dissected, the Vesalius illustration represents a theatre crowded with not only doctors but also with hundreds of spectators: women, working men and even a dog can be found in the packed audience, its lively public air in stark contrast to Rembrandt's restrained professional one.

Von Hagens' own exhibitions, as popular as they are controversial (in Vienna, for instance, the exhibition was kept open 24 hours a day to cope with public demand), belong as much to this rather less exalted heritage of popular anatomical theatres and the anatomical museums of the eighteenth and nineteenth centuries examined in the previous chapters as they do to the world of professional anatomy represented in the form of Dr Tulp. Moreover, von Hagens' business and promotional practices bear a striking resemblance to those of the proprietors of commercial anatomical museums. Like Barnum before him, von Hagens is a tireless self-promoter. He features prominently within the space of his own exhibition, representing a stylised image of himself—always in his trademark black fedora—as part of his own displays. Publicity material often depicts von Hagens playing chess with the "Chess Player," or facing a plastinate garbed in a matching black hat, or standing between the halves of a vertically sliced exhibit. As we have seen, von Hagens, like Barnum, recognises the commercial benefits of controversy, and of public debate about the meaning of the objects of display and the nature of the exhibition itself.

In this way, von Hagens' *Body Worlds* exemplifies the key characteristic of the all-anatomical spectacles examined throughout this book, moving constantly between the popular and the professional, science and art, education and entertainment, both problematising and reinforcing the relationship between them. Since the early modern period, as we have seen, anatomy and spectacles have been mutually constitutive—sometimes productively intersecting and sometimes defining themselves against one another. Rembrandt's 1632 painting *The Anatomy Lesson of Dr Nicholas Tulp* and the cover image of Vesalius's *De Humani Corporis Fabrica* represent two very different ways of seeing the act of anatomical dissection in public places—one sober and professional, the other a crowded circus—and so draw attention to the multiple possibilities and sometimes contradictory conventions encoded not only in anatomical modelling, in the way anatomy sees the body, but in the way anatomy sees itself and its own practice.

The Anatomisation of "Real Human Bodies"

Despite their differences, a key component of all of the exhibitions we have been examining throughout this book has been their representation of the anatomical gaze as that which allows us to discover the truth of the body. Von Hagens' work, which constantly urges the viewer to "look inside" the body, is in many ways the apotheosis of this tradition. The press release for the 2006 *Body Worlds* exhibition at the Denver Museum of Nature and Science typifies this, encouraging visitors to "Uncover the incredible beauty and complexities beneath your skin" (Denver exhibition flyer), while an early exhibition video for *Body Worlds* questions asks, in a similar vein: "What is concealed beneath the skin which protects and covers our body?" (cited in Wegenstein 227–28). An advertising postcard for the 2006–2007 Philadelphia exhibition features an image of a woman's back with the words "look inside" printed on a detachable panel, which can be lifted to show the anatomy of the spine. This emphasis on revealing something that is usually concealed is central to the show's advertising rhetoric and to its curatorial practice.[9] By making visible the body's internal anatomy and its function, *Body Worlds* repeatedly claims, its écorché figures and partial body plastinates show the true condition of its health, thereby reinforcing the exhibition's stated objective to teach the public about the importance of anatomical knowledge to the (self-)cultivation of good health. Comparative pathology exhibits—such as "Smokers' Lung," displayed next to a healthy specimen—are explained, in their accompanying text, as revealing the "truth" of an internally diseased organ whose deleterious effects are not necessarily visible on the surface of the skin. "Shrunken Liver," similarly, places a cirrhotic liver next to an unaffected one, in order to caution against an "excessive" alcohol consumption whose ravages, again, might be wrought inside the body rather than outside. In a new exhibit shown for the first time in Philadelphia, "Obesity Revealed," two vertically sliced bodies (one weighing 580 lbs and the other 140 lbs) were displayed side by side, in order to demonstrate the harmful effects of "excessive" amounts of fat on the vascular system. The logic behind these exhibits is that behaviours such as smoking, drinking and fat consumption have dangerous consequences whose effects might not be evident on the surface of the body. Thus the body must be opened and the internal anatomy scrutinised in order to assess the real state of that body's health. At the same time, however, such exhibitions do not simply reveal or make visible the condition of these organs in an unmediated way. Rather, these are artificial constructs that have been dyed and treated to simulate the visible signs of the diseases they are used to represent.

In this way, the "naturalness" and "transparency" of von Hagens' plastinates are wholly constructed characteristics: these are bodies in which the fluids have been replaced by synthetics, the organs dyed to render them more clearly visible, and whose poses are maintained by the insertion of metal pins. The écorché figures *Body Worlds* advertises as "real human bodies" are neither wholly real (in the sense that they are synthetic creations), nor "human" (in the sense that 70 per cent of their mass has been replaced by plastic). As Megan Stern argues:

> Ironically, given von Hagens' insistence that ordinary people should have access to 'the real thing', these are 'actual' human bodies that have been reworked to look like aesthetic representations of the body. The plastination process itself, in which all bodily fluids are replaced with a variety of synthetic materials including silicone rubber, epoxy resin and polyester, enhances this sense of the bodies as constructs. The plastinate, denuded of the qualities that would make it fleshly, becomes a static, odourless, impermeable and clearly delineated reworking of the original body. (84)

That von Hagens defines the "real" body in such a particular way — while simultaneously claiming to reveal its unmediated "truth" outside all systems of interpretation — is exemplary of the tendency we have been seeing throughout this book, in which anatomy repeatedly disavows and distances itself from modes of visualisation that are, nonetheless, inherent to its production and circulation of knowledge about the body.

The disparity between von Hagens' rhetoric about the unmediated reality of the bodies exhibited in *Body Worlds* and his actual practice of anatomical modelling, which offers a highly constructed and conventional representation of those bodies, is again exemplified by the play between "real" and "copy" that takes place between the "Skin Man" plastinate and the "Muscle Man" illustration. While von Hagens clearly intends his plastinates to be seen not as copies but as the triumphant materialisation of the Renaissance écorchés—the "real" version of the body that Valverde's illustration only represents—they are also, as Van Dijck recognises, not a return to the real but second-order images, "*imitations of representations*" (114; original emphasis). In the juxtaposition of "Skin Man" and "Muscle Man" we might therefore read the "real" body of von Hagens' plastinate as a simulacrum of the image which inspired it. The *Body Worlds* exhibits might thus be seen not as the fulfilment of a long tradition of anatomical art through the use of modern technologies like plastination, but as a de-realisation of the body's materiality and its transformation into the image of an image. Despite its claims to put this material, original body on display, what *Body Worlds* actually shows is the extent to which these ideas about "real bodies" are themselves cultural constructs, produced by the wider conventions within which we have learned to see and know the body.

The conventionality encoded in *Body Worlds'* modelling of its plastinates, and the extent to which this problematises its claims to represent the body itself ("free of interpretation") is strikingly evident in von Hagens' careful distinction between the kind of "real body" his plastinates represent—preserved whole-body anatomical specimens—and the reality of the dead body or corpse. *Body Worlds* is not an encounter with death, von Hagens insists: the exhibition is not a "post-mortem museum," he emphasises. "This is not a place for mourning. It is not an illegal cemetery—it is a hall of enlightenment and when you need to learn you cannot mourn" (www.bodyworlds.com). In this way, *Body Worlds* reproduces exactly the same distinction made by the early anatomical modellers Chovet and Desnoües in the first chapter of this book, who wanted to provide information about the body's anatomy to a popular audience "without exciting the feeling of horror men usually have on seeing corpses" (Desnoües, quoted in Haviland and Parish 56). In his 2002 interview with the *Guardian*, von Hagens similarly explained: "I want to bring the life back to anatomy. I am making the dead lifeful again" (Jeffries 3). In her analysis of von Hagens' work, Megan Stern argues that the cultural acceptability of *Body Worlds* is dependent on its disassociation from actual, or "real," death:

> Real, decaying corpses are messy and smelly, qualities which play a crucial role in rendering the corpse taboo, the destabilizing abject that must be made safe through rituals of purification and detachment.

> Plastination arguably constitutes just such a ritual, so that instead of shocking visitors by confronting them with abject corpses, 'Body Worlds' renders these corpses safe, unthreatening. 'Body Worlds' is reassuring because, whilst undoubtedly promising the ghoulish thrill of encountering authentic human corpses, it also neutralizes this encounter. It gives us the corpse in spectacular fairground mode: exciting but safe. (88)

The dead body, and the putrefaction of the corpse, are precisely the parts of the body's "reality" that are excluded during the process of plastination and disavowed within the space of the *Body Worlds* exhibition. In this respect, as Stern recognises, *Body Worlds* does not, despite its advertising, represent "real" bodies, unmediated by interpretative frameworks: "The original body cannot anywhere show itself because it constitutes the very material from which the simulation is made. The actual body has been displaced by a hyperreal one whose durability and authenticity entirely displace the need for the real thing" (87).

Moreover, just as *Body Worlds'* exhibition of "real bodies" serves to construct the bodily reality it claims to reveal, so does its curatorial focus on the écorché figure and its exposed internal anatomy serve to construct rather than reveal the bodily interiorities put on display. In this, again, von Hagens reflects the legacy of the early modern anatomical knowledge he has inherited, and which he publicises himself as single-handedly invigorating: specifically, that the "truth" of the body's health is to be found in the exposure and examination of its interiority. This idea, too, is one that has its origins in the early modern anatomical tradition von Hagens' so explicitly references, which is both exemplified and embodied by the emergence of the écorché as a privileged signifier of anatomical science during this time. As such, écorché figures both reflect and produce an epochal shift in the way that bodies were conceptualised, which has continued to have a profound, though rarely recognised, influence on contemporary understandings of embodiment. Thus, what *Body Worlds* reveals is the ongoing influence of the early modern anatomical tradition he cites on twenty-first century understandings of bodies, reflecting a continuity between the emergence of an anatomical vision of the body, as seen in the first chapters of this book, and that of the present day.

In order to understand how the rise of practical anatomy as an increasingly central part of medical practice could so profoundly transform the way bodies were previously understood—giving rise to a specifically modern concept of both the body and the self-cultivating subject, as tracked throughout this book—we need to understand how it differed from the conceptualisation of the body that precedes it. The

modern concept of the body as autonomous and highly individuated
emerged only in the seventeenth century; before this time, a medieval
concept of the body prevailed, in which it was understood as "a
flexible, fluid quantity" (Austin 41–42), one that was constantly "dissolving and reconfiguring its own boundaries" (Perry 146). Prior to the
early modern period, Daniela Bohde argues, people "did not consider
themselves to be isolated," or individuated, but rather saw the body
as open and unstable in a way reflected, for instance, in theories
about the spread of plague, in which "pestilence penetrated through
openings of the skin—eyes, mouth and nose, but also the pores"
(31). Over the course of the late sixteenth and seventeenth centuries,
however, this medieval "perception of the body as porous, open, and at
the same time interwoven with the world" came to be replaced "with
one that viewed it as an individuated, monadic, and bourgeois vessel
that the subject was considered to inhabit" (Benthien 37). This is the
shift Bakhtin has influentially identified as one from a "grotesque"
to a "classic" bodily canon. Bakhtin writes that the medieval schema
that conceived of the body as protruding into and intermingling with
the world around it gave way during the early modern period to "an
entirely finished, completed, strictly limited body, which is shown from
the outside as something individual. That which protrudes, bulges,
sprouts, or branches off (when a body transgresses its limits and a new
one begins) is eliminated, hidden, or moderated. All the orifices of the
body are closed" (320).[10]

It is not coincidental that practical anatomy—and subsequent
prevalence of images of the opened, flayed body within anatomical
art—should emerge contemporaneously with this new conceptualisation of the body as self-enclosed, bounded and individuated. Rather,
anatomy's focus on opening up the bodily interiority and a new understanding of the body as something bounded and self-contained are
mutually dependent developments, (re)producing a new distinction
between the inside and the outside of the body. The skin takes on
a heightened cultural significance at this time for just this reason,
and representations of its removal particularly so: it is through this
new understanding of the skin as the boundary of an individuated
body (rather than a porous membrane) that the very concept of a
bodily interiority becomes meaningful, or even possible. The idea of
a concealed or hidden bodily interior is thus established through the
act of representing its exposure. In this way, early modern écorché
illustrations do not simply represent the peeling away of the body's
outside in order to reveal its inside but, on the contrary, represent the
establishment of the skin as a border of an individuated self.

In keeping with the interdependence between medical and spectacular cultures we have been examining throughout this book, it is
important to recognise that this change in the way of seeing the body
in the seventeenth and eighteenth centuries (re)produced concurrent
changes taking place in wider systems of knowledge. This is not to say
that this new way of seeing and understanding the body is simply a
product of the new forms of knowledge emergent during this period:
rather, as we have seen in the previous chapters, modes of visualisation,
cultural imaginaries of the body and medical knowledge are mutually
constitutive and mutually (trans)formative. Thus the privileging of
the bodily interior in anatomical art reflects, as Petra Kuppers writes,
a more general "shift of observational emphasis from the outside to
the inside" at this time, which in turn produced a "different paradigm
of medical knowledge" ("Visions of Anatomy" 124). The significance
attributed to the removal of the body's skin in écorché figures exemplifies this, and is characteristic, as Claudia Benthien argues, of the way
that "Western thought since the Renaissance has been dominated by
the fundamental notion that knowledge of what is essential means
breaking through shells and walls in order to reach the core that lies
in the innermost depths," a view reflected in "the anatomical paradigm
of penetration and uncovering that was established in the sixteenth
century by Andreas Vesalius" (7). In this schema, "skin is imagined
as a protective and sheltering cover but … also as a concealing and
deceptive one" (17). For Barbara Stafford, too, early modern anatomy
understood as the rendering visible "hidden morphologies 'in spite of
the skin which veiled them'" (54). The work of the anatomical artist
was thus to expose and examine the "forms concealed beneath an
occluding matter" (54).[11] It is for this reason, as noted in the introduction to this book, that anatomy is not only a field of knowledge about
the body but also a way of seeing the body. Anatomy itself, then, is a
kind of visualising technology.

If the status of the "real" body is a good deal more complex than
von Hagens, with his claims that his plastinates are not models but
bodies exhibited "free of interpretation," suggests, it is precisely to
this complex nature of their signification that we can attribute the
immense popularity of his work. That is, although von Hagens makes
explicit the logic of the anatomical vision and reflects its dominance as
a way of seeing the body, it is also possible that audiences are drawn to
these exhibitions precisely because they are sites in which the natural
and the artificial, the scientific and the freak show, the inside and the
outside, are both revealed and destabilised, in a cultural context in
which contemporary ideas about bodies have again been challenged by

the interventions of modern medical technologies. The idea of the "real body" itself has itself become increasingly precarious in the twenty-first century. The rapid proliferation of medical biotechnologies — from common practices such as the use of glasses and dentistry, to cosmetic procedures like botox and rhinoplasty, to major surgical procedures like artificial hips and organ transplants — means that the twenty-first century body is rarely entirely "natural," and assumptions about what constitutes a real body have accordingly needed to be rethought.

Despite the conventionality of *Body Worlds'* modelling, and its explicit reinforcement of normative ideas about the body and its care (such as the need for self-moderation and self-regulation), these plastinates may well be compelling not because they offer reassuring advice about how to treat the body, but because we live in a period in which these bodily norms are fraught and failing, transformed by medical biotechnologies and new media (of which plastination itself is one example). The removal of the body's limits to make these écorché figures is fascinating in a context in which the limits of the body are increasingly uncertain and problematic. To return to the contrast between Valverde's écorché "Muscle Man" and von Hagen's plastinate "Skin Man," the continuity between illustration and embodiment staged in and by *Body Worlds'* écorché figures also reflects the extent to which anatomical art is, to recontextualise Judith Butler's words, "that which exceeds and compels that signifying chain, that reiteration of difference" (79). In a similar way, although, as we have already seen, von Hagens repeatedly represents the significance of his plastinates as revealing to the public the hidden, unseen interiority of the body — as fulfilling the anatomical vision of the early modern period, in which the body's skin, seen as a barrier to knowledge, has been removed and the body's interiority rendered perfectly visible — I would argue that the *Body Worlds* exhibits instead reflect how problematic and illusory the idea of the "body's interiority" has always been within anatomy. Below the surface of the skin, von Hagens' displays reveal, what we find is not its unproblematic truth laid bare, but rather a series of further surfaces. His vertical slices exemplify this. These are thin dissected cross-sections hung separately, so that the body can be seen through in a way not possible with an intact écorché figure, but which also transforms the deep interiority into a form of exteriority. The body's interiority can only be represented when brought to — and depicted as — the surface.

Rather than see the history of anatomical art as one of an increasing realism, in which the body is represented entirely objectively, "like a medical textbook," we might see it as one in which a new idea of realism and a new understanding of the real body emerges. Scientific knowledge of anatomy is not simply becoming more detailed and objectively better, but is, rather, increasingly codified, to ever more microscopic levels, as the material or "real" body recedes ever further from view. This disappearing body, always threatening to vanish, is, paradoxically, not becoming increasing invisible, not disappearing from the public sphere, but, on the contrary, experiencing a period of heightened popular interest, circulating through popular media and sites of spectacle. Many forms of mass media — television, film, advertising, tabloid magazines, popular literature and the Internet — continue to circulate anatomised images of bodies, both to inspire self-improvement and to warn about the danger of failing to exercise self-discipline and to guard one's health. In this way, *Body Worlds* is exemplary of the history we have been examining throughout this book, in that it transforms the tradition it continues, problematises the body it puts on display, and provides a site — as did all the exhibitions we have examined in the previous chapters — in which spectators can learn how to see and understand their own bodies in contexts that both confirm and challenge normative assumptions about them.

Notes

1 It is uncertain, however, whether Wilson's protest was a genuine criticism of *Body Worlds* or merely a staged one. Wilson was able to mount one of the exhibits — an écorché horse — and, despite the fact that photography is prohibited at the *Body Worlds* exhibitions, a photograph of Wilson's naked protest later appeared in von Hagens' own *Body Worlds Magazine*, titled "Lady Cadaver." Moreover, given the length of time to produce each plastinate and the speed with which female plastinates in active poses have since appeared on display, there are very real questions about whether the whole protest was simply a stunt (MacDonald 7).

2 The *Body Worlds* exhibitions currently on display are part of *Body Worlds 3*, whose content reflects "a return to the Renaissance [and] Middle Ages" (www. bodyworlds.com). The earlier shows were the original *Body Worlds*, "inspired by the anatomists and anatomical drawings of the Renaissance" and *Body Worlds 2*, which is more "dynamic because it offers more sportive poses [...] for example, plastinates performing yoga, or rollerblading, or about to perform a dive." The current shows are described as part of an interconnected suite known as "The Human Saga," and said to be "[i]nspired by the latest advances in neuroscience, cardiology, biology, genetics, gerontology, psychiatry, and physiology" (www. bodyworlds.com).

3 As the UK Anatomy Act is applicable only to British bodies, or those deceased in the UK, von Hagens appears to have successfully exploited a loophole in British law by the simple expedient of importing the body used in the autopsy from Germany.

4 This claim is somewhat undermined by the high ticket prices for the show:

usually around $US30. In Philadelphia, with its multiracial population, almost all the spectators were white.

5 Von Hagens' exhibitions continue to have a strong commercial element, with an extensive range of exhibition merchandise—posters, puzzles, key rings, mouse pads and children's toys—available for purchase.

6 *Body Worlds* is still often evaluated according to whether it "succeeds" as an educational exhibition. See, for instance, recent studies by Walter, vom Lehn, and Brown.

7 Von Hagens claims to receive thousands of enquiries about body donation during every show, and even shows that do not accept donations—like *The Amazing Human Body* in Australia—claim to be flooded with offers after each exhibition. There are no publicly available records for the number of body donation forms actually completed and submitted, however. Given that the donation agreement requires the estate of the deceased person to meet all shipping costs to Germany, however, one wonders what percentage of bodies are actually acquired by this means.

8 By far the single most contentious issue surrounding *Body Worlds* and other such exhibitions is the provenance of the bodies they put on display. In 2005, for instance, the UK newspaper The *Telegraph* reported that von Hagens had been forced to remove his father, a former SS officer, from his role as project leader "at a new corpse preparation factory in Poland" (Connolly 2005). Critics of *Body Worlds*, argues Helen MacDonald, question: "how [von Hagens] obtains so many bodies. They note that the plants at which they are processed are located in remote places thought to lack ethical credentials where human remains are concerned. One is in China, where von Hagens now lives. Another, a medical institute in Novosibirsk, Russia, ships him suspiciously large numbers of bodies, from people who were mentally ill, destitute, or had been imprisoned" (5–6).

9 While museums like the Hunterian and Wellcome in London and the Mütter in Philadelphia contain human remains and are open to the public, no new remains have been added to their public collections in well over a century. The Wellcome opens only its Museum of the History of Medicine to the public; the Museum of Anatomy and Pathology contains specimens, including wet tissue specimens, that are less than ten years old, and access is legally restricted to members or students of the medical profession. The National Museum of Health and Medicine in Washington DC similarly exhibits to the public only historical remains (jncluding preserved limbs amputated during the American Civil War), while an extensive collection of twentieth-century remains is held in closed archives opened only to medical professionals.

10 At the same time, as Daniela Bohde cautions, it is unlikely such transforma-tions took place as "a hard and definite change in paradigm, from an open to a closed body type," noting that, "[i]t is far more likely that the various forms of understanding the self existed simultaneously" (32).

11 It is this view, Stafford notes (84), that encourages the eighteenth- and nineteenth-century interest in physiognomy and phrenology, in which physical signs were seen as "indicative of hidden causes legible only to specialised interpreters." Stafford argues: "The master eighteenth-century physiognomist, Lavater, noted that men form conjectures 'by reasoning from the exterior to the interior' [...] The 'science' of physiognomies was predicated on the ability to draw inferences from 'known' surfaces to unknown depths, and from parts to wholes" (107).

Conclusion

This book has examined the way anatomised representations of bodies have become objects of public spectacle since the simultaneous rise during the early eighteenth century of practical anatomy and of the modern exhibitory culture, as well as the way particular bodies become popular as objects of such spectacles during periods when their status and significance are undergoing rapid cultural change. As we have seen in the examples of the maternal body, the sexual male body, the disabled body and the plastinated body, examined in the preceding chapters, throughout the modern period sites of public spectacle featuring anatomical exhibitions of human bodies have provided, and continue to provide, a space in which the meaning of the bodies on display can be debated, and the transformations to which they are subject can be visualised and interpreted. Although it has been widely assumed that the history of such exhibitions is one in which medical discourses gradually came to usurp and replace popular ones, this book has shown that spectacular and scientific ways of seeing the body have continued to be closely interrelated and mutually dependent. Certainly, medical discourses have enjoyed an increasingly dominant position over the course of the last three centuries as a privileged source of knowledge about the body and its care; however, as we have seen, medical institutions established themselves in this role, and in the popular imagination, through recourse to modes of spectacularisation that are thus better understood as intrinsic to medical knowledge (that is, part of the process by which this knowledge is constituted) rather than extrinsic to it (that is, simply a vehicle by which medical knowledge is [mis]communicated).

The kinds of bodies, the sites of public spectacle and the types of knowledge featured in anatomical exhibitions have varied widely across the period from the early eighteenth century to the present, but the popular fascination with such exhibitions, and the distinctive combination of the medical and the spectacular on which they draw, have remained fairly constant (although they have waxed and waned over time, as we have seen). The cultural space mapped out at the intersection of the medical and the spectacular, between the technologies of the image and the technologies of the body, has continued to be an intensely productive one, providing a highly influential site at which contemporary ideas about bodies and subjectivities are formed and transformed in ways that have had a demonstrable impact upon the public sphere. The case studies examined in this book, while not intended to provide an exhaustive account of the history of public anatomical exhibitions, have together demonstrated that such exhibitions shared a central characteristic in providing an important locus at which changing ideas about bodies could be displayed and debated. Thus, the exhibitions examined in the preceding chapters do not simply represent bodies whose meanings are already fixed and determined, but are sites in which new bodies are actively constructed. In this respect, the history of anatomical exhibitions is exemplary of the history of the body itself as described by Michael Feher. Feher argues that the history of the body is "neither a history of scientific knowledge about the body nor a history of the ideologies that (mis)represent the body. Rather, it is a history of 'bodybuilding,' of the different modes of construction of the human body" (159). It is just this kind of "bodybuilding" one finds in the history of public spectacles featuring displays of human anatomy. While the history of practical anatomy might seem highly specialised, of most interest to narrow histories of the medical profession, and while anatomical exhibitions might appear to have occupied a marginal role within the history of popular culture, as we have seen in this book, they have nonetheless had a profound impact on ideas about bodies in the popular sphere and in establishing the widespread importance of anatomy for the general public.

This is particularly evident when we consider the role of public spectacles and anatomical exhibitions in promoting what has now become a firmly entrenched way of understanding and relating to the body, in which its condition—both its health and its appearance—is seen to be largely under our individual control and thus a matter of personal responsibility to cultivate. As this book has shown, anatomical exhibitions have repeatedly served to popularise the importance of undertaking practices of self-care as a compulsory form of work on the self. As the interiority of the body has become ever more visible to systems of knowledge and power, as it has come increasingly under the material reach of modern medicine, so has the significance of the exteriority of the body been correspondingly transformed in the popular imagination, read as a measure and reflection of our character, temperament and the disciplinary practices we undertake on ourselves. The body, in the modern period, is increasingly viewed as a legible text on which can be read the signs of our personal conduct and private habits. The anatomical interiority of the body is hence constructed as a hidden truth that must be exposed to the penetrating gaze of power and as something that inevitably inscribes itself on the surface of the body. Modifying Feher's idea of "bodybuilding," then, we might see the history of anatomical exhibitions not as a space in which we see the making of the modern body (to paraphrase Gallagher and Laqueur) but rather as one that traces the emergence of the modern idea of the body as self-made, as something we are each individually required to craft and cultivate.

This approach towards and understanding of the body has only intensified in the late twentieth and early twenty-first centuries, a period in which, not coincidentally, public spectacles and exhibitions featuring anatomised representations of human bodies have again become highly popular. This is evidenced not only by the continuation of live public spectacles like von Hagens' *Body Worlds*, examined in the previous chapter, but also in the vast wave of television shows, from *Embarassing Illnesses* to *Extreme Makeover* to *You Are What You Eat*, which routinely feature highly spectacularised anatomical images of the body's interiority. The popular series *The Biggest Loser* is exemplary of such shows and the extent to which they both continue and transform the tradition of live exhibitions examined in this book. Catering to the same fascination with non-normative bodies that fuelled the popularity of the nineteenth-century museums of anatomy and freak shows, in which professional fat men and women were enduringly popular attractions, the competitive weight-loss show *The Biggest Loser* draws many of its modes of spectacularisation from these earlier exhibitory spaces. It emphasises the prodigious weight of its competitors (relying particularly on highly dramatised public weigh-ins to which, especially in the first week, the contestants are shown to have strong emotional responses) and spectacularises their eating habits, as well as their physical size. At the same time, the representation of the competitors' bodies in this show is also consistent with the increasing medicalisation of the body we have seen over the modern period. At the start of each series, contestants are subjected to a series of medical tests said to reveal the damage obesity causes the body: the high blood pressure, incipient diabetes, strain on the heart, and reduced lung capacity that undermine the body's health. The competitors' "anatomical age" (that is, the "true" condition of their bodily interiority) is measured and compared to their chronological age.

In this way, the generic conventions of *The Biggest Loser* reflect the increasing dominance of medical institutions and discourses in the way bodies are represented in the popular as well as the professional spheres, while reinforcing the cultural emphasis on our personal responsibility for the condition of our bodies. The contestants are never diagnosed with specific anatomical or genetic conditions, and neither are cultural conditions, such as work and family require-ments, considered contributing factors to their size. Instead, this is always attributed to personal diet and lifestyle choices; that dif-ferent bodies are inclined to be different sizes is a possibility never canvassed. *The Biggest Loser* is thus exemplary of the way the medical becomes the moral, providing a legitimising discourse through which to justify normative assumptions about bodies and to shame or to motivate spectators, as well as participants, about their treatment of their own bodies. *The Biggest Loser*'s disciplining of the contestants' bodies, often in punishing and extreme ways—such as dramatically restrictive diets or gruelling work-outs that leave the contestants weeping or vomiting—is reflective of the cultural disapproval that accrues around failures to care for the body in normative ways, and the severity with which apparently undisciplined subjects can be forced to assume responsibility for their bodily condition. (We see here echoes of the harsh treatment of the spermatorrhoeaic body examined in Chapter 2.) Moreover, the contestants on *The Biggest Loser* are required to internalise the advice provided by the show's hosts and physical trainers; these are provoked in the course of tearful con-fessions during which the contestants are encouraged to talk about their relationships to food and their bodies in highly psychologised terms, and they invariably end with each resolving to take better care of his/her future health and weight.

A popular television show like *The Biggest Loser* is an illustrative example to consider at the conclusion of this history of anatom-ical exhibitions because it demonstrates the extent to which, while the fact of the public fascination with spectacularised representations of unusual anatomies has remained constant over the course of the modern era, the interpretative frameworks and cultural narratives within whose contexts these are understood have changed substan-tially. As both Alice Dreger and José van Dijck have noted (in *One of Us* and *The Transparent Body*, respectively), representations of conjoined twins remain popular in contemporary media: they are featured in high-profile films and television shows, documentaries and newspaper reports, in many ways continuous with those found in the nineteenth-century freak show exhibitions of twins such as Chang and Eng (the original Siamese twins), Millie-Christine (African-American twins) and the Hilton Sisters (who were also musical performers). Yet the contemporary media invariably focuses its representations on separation surgery, providing a medicalised framework for depictions of conjoinment in which the emphasis is on their treatment and potential cure. A similar tendency can be seen in *The Biggest Loser*, with its tracking of the contestants' weight loss and their psychological as well as their physical transformations, which justifies the show's spectacularisation of the fat body by focusing on its medical improvement and physical normalisation. The competitors are shown becoming progressively healthier and happier as their size diminishes week by week, although recently at least one former contestant has confessed that she developed an eating disorder as a result of the show, and claimed that the show's staff encouraged practices such as dehydration before weigh-ins in order to maximise dramatic results. In this way, while the twenty-first century culture of television shows and tabloid magazines remains fascinated with the kinds of bodies featured in nineteenth-century anatomical exhibitions and public spectacles, such spectacularisation is considered acceptable and uncontroversial only so long at its ostensible focus is on the capacity of modern medicine to cure or at least help such bodies.

As medical technologies play an increasingly prominent role in the care and treatment of the body, expanding into an ever-widening range of practices—from minor procedures such as laser eye treatment or tooth-bleaching to elective surgeries such as rhinoplasty or stomach-banding, diagnostic services such as digital imaging and scans, and major medical treatments such as chemotherapy and transplants—the body has come to be seen as increasingly more malleable, infinitely perfectible. Contemporary culture is thus one in which practices of self-improvement are simultaneously represented as disciplinary technologies, as well as privileged forms of self-expression. In such a context, in which medicine has transformed popular assumptions about the limits of the body and its capacity to be reshaped, it is unsurprising that spectacularised representations of the body's anatomy have again become the object of fascination. Images of disastrous cosmetic surgery, advertisements for medical procedures that promise dramatic results, cultures of extreme body modification and the practice of artists such as Orlan and Stelarc are all examples of this. In *Body Invaders*, Arthur and Marilouise Kroker question:

If, today, there can be such an intense fascination with the fate of the body, might this not be because the body no longer exists? For we live under the dark sign of Foucault's prophecy that the bourgeois body is a descent into the empty site of a dissociated ego, a "volume in disintegration," traced by language, lacerated by ideology, and invaded by the relational circuitry of the field of postmodern power. (20)

Baudrillard contends that we are living in a hyperreal culture in which the assumption that there is a clear distinction between original and copy, real and fake, the object and the representation has dissolved into a precession of simulacra. Recently, theorists have similarly argued that traditional concepts of the body as a natural, self-contained and individual entity have given way in the twenty-first century to a new view of the body as technologised, artificial and post-human. In Rosi Braidotti's description of this new body:

the post-human body is not merely split or knotted or in process: it is shot through with technologically-mediated social relations. It has undergone a meta(l)morphosis and is now positioned in the spaces in between the traditional dichotomies, including the body-machine binary opposition. In other words, it has become historically, scientifically and culturally impossible to distinguish bodies from their technologically-mediated extensions. (228)

For Braidotti, meta(l)morphosed bodies open up a space for possibilities and potential becomings by problematising the modern concept of the body as individual, self-contained and unchanging in a way that returns us to a pre-modern understanding of the body (noted in Chapter 3) as fundamentally open to and interrelated with the other bodies around it. For Beatriz Preciado, similarly, the early twenty-first century has already distinguished itself by the "assemblage of new technologies of the body (biotechnology, surgery, endocrinology ...) and of representation (photography, cinema, television, cybernetics ...) which infiltrate and penetrate life as never before" (74). The result, she argues, is a contemporary body that "is not a passive living material but a techno-organic interface, a techno-living system segmented and territorialised by different political models (textual, computerised, biochemical)" (108). In the words of the Critical Art Ensemble, the contemporary body is widely understood as a "flesh machine," a "heavily funded liquid network of scientific and medical institutions with knowledge specializations in genetics, cell biology, biochemistry, human reproduction, neurology, pharmacology, etc., combined with nomadic technocracies of interior vision and surgical development" (4).

As the rapid proliferation of bio-technologies challenges what we think a body is and what it can do, what we see, on the one hand, is the development of a vast array of cultural institutions designed to enable subjects to undertake the self-making of their bodies (from private medical clinics to self-help books to television make-over programs), while, on the other hand, there is also increasing public concern and government regulation surrounding reproductive technologies, commercial bio-banking and research involving the use of stem-cell material. To paraphrase de Beauvoir, the twenty-first century body is made, not born. What we have seen in this book, however, is that while contemporary medical technologies do have an increased capacity to materially redesign the body in a way specific to the twentieth and twenty-first centuries, the knowledge frameworks through which bodies are understood, the lenses through which they are seen, themselves constitute a form of technology that has always shaped our understanding of bodies and our experiences of our own bodies. That is, and as we have seen in the examples of the anatomical exhibitions examined in the preceding chapters, anatomical science does not merely rely on medical and visual technologies as the tools with which to dissect bodies or to represent the knowledge acquired by that dissection; rather, anatomy itself needs to be understood as a form of technology, one that has played, and continues to play, a central role in constituting popular ideas about embodiment and in shaping public attitudes towards the body and its care. For this reason, the technological transformation of the body, the "building" of bodies through modern technologies like that of medicine, should not be seen as a new development, the distinguishing feature of the post-human, or postmodern, body. Rather, it is a characteristic of all bodybuilding, of the construction of bodies, that not only their cultural significance but also the way they are experienced (the production of their interiorities as well as exteriorities) are products of their technologisation; that is to say, of their inscription in and by systems of knowledge and their interpretation through particular ways of seeing. The recognition of anatomy as one such technology is especially important given that, as we have seen, anatomy continues to enjoy an unusually privileged status as a source of direct and unmediated truth about the body. This book, in positioning the history of practical anatomy within the history of popular culture, has demonstrated how the ongoing relationship between medical and spectacular cultures has continued to shape what we think a body is and what it can do, in ways whose origins are rarely acknowledged.

Bibliography

Exhibition Catalogues, Handbills and Other Promotional Material

Alison, J. (Surgeon). *A Description, Anatomical and Physiological, of the Sectional Model of the Human Body, The Parisian Venus; Comprising a Popular Account of the Parts Displayed: The Function and Uses*. "Know Thyself." Manchester, UK: A. Burgess, 1844.

The Amazing Human Body. Exhibition pamphlet. Sydney, Australia 2006.

Bailey, J. A. and P. T. Barnum. *Barnum and Bailey Official Route Book: Season of 1893*. Buffalo, NY: H.L. Watkins, 1893.

Barnum, P. T. *How I Made Millions: or, The Secret of Success*. Chicago, IL, and New York: Belford, Clark, 1884.

Body Worlds. Exhibition pamphlet. Philadelphia, PA 2006.

—. Website: www.bodyworlds.com.

A Catalogue and Description of the Human Anatomy in Wax-Work, and Several Other Preparations, to be Seen at the Royal Exchange. London. 1736.

The Celebrated Florentine Anatomical Venus: Together With Numerous Smaller Models of Special Interest to Ladies, Showing the Marvellous Mechanism of the Human Body. Huddersfield, UK: n. p., n. d.

Coney Island website: www.coneyisland.com.

Critical Art Ensemble. *Flesh Machine: Cyborgs, Designer Babies, and New Eugenic Consciousness*. New York: Brooklyn Autonomedia, 1998.

A Curious Collection of Prodigies, Dwarfs, Giants, Aged People, Twins, Extraordinary Animals, Monstrosities, Wonderful Freaks of Nature, Mermaids, & c & c & c. Ashmolean Hope Collection Folio B 29. Oxford: n. p., n. d.

Curious Exhibition Bills of Giants, Dwarfs, Monstrosities & c & c. Ashmolean Hope Collection Folio B 29. Oxford: n. p., n. d.

A Descriptive Catalogue (Giving a full Explanation) of Rackstrow's Museum: Consisting of a large, and very valuable Collection, of most curious Anatomical Figures, and real preparations: also Figures resembling Life; With a great Variety of Natural and Artificial Curiosities. To be seen At No. 197, Fleet-Street, between Chancery-Lane and Temple-Bar. London: n. p., 1784.

An Explanation of the Figure of Anatomy Wherein the Circulation of the Blood is Made Visible Through Glass Veins and Arteries, with the Actions of the Heart and Lungs; As also, the

Course of the Blood from the Mother to the Child, and from the Child to the Mother. London, 1737.

Fraser, Mat. Website: http://www.matfraser.co.uk.

Grand Musée Anatomique et Ethnologique du Dr P. Spitzner. Ethnologie–Anatomie Humaine, Pathologie, Chirurgie et Chirurgie Obstétricale, Teratology, Vénérolgie. [Auction catalogue]. Paris: n. p., 1985: n. pag.

Happy Sideshow. Website: http://www.thehappysideshow.com.

"International Health Exhibition, London 1884." *The Health Exhibition Literature*. Vol. 10. London: William Clowes, 1884.

Jordan, Dr Louis. *The Philosophy of Marriage. Being Four Important Lectures on the Functions and Disorders of the Nervous System and Reproductive Organs.* In *The Handbook of the Pacific Museum of Anatomy and Natural Science*. San Francisco, CA: n. p., 1865.

Jordan, Robert Jacob. *The Handbook of the Pacific Museum of Anatomy and Natural Science*. San Francisco, CA: n. p., 1865.

—. *The Illustrated and Descriptive Catalogue of the Subjects contained in the London Anatomical Museum to which is annexed the Guide to Masculine Vigour*. London: 1892.

Jordan and Davieson, Drs. *Practical Observations on Nervous Debility and Physical Exhaustion: to Which is Added an Essay on Marriage, with Important Chapters on Disorders of the Reproduction Organs. Being a Synopsis of Lectures Delivered at the Museum of Anatomy, Philadelphia*. Philadelphia, PA: n.p., 1871.

Kahn, Joseph. *Handbook of Dr Kahn's Museum*. London: W. Snell, 1863.

Morning Chronicle, 17 April 1824: 3.

A Most Strange and True Report of a Monstrous Fish, that Appeared in the Form of a Woman, from her Waist Upwards Scene in the Sea by Divers Men of Good Reputation. London, 1604.

Our Body: à corps ouvert. Exhibition pamphlet. Paris 2009.

Philadelphia Museum Company Minutes and Records, 1808–1842. Smithsonian Institution Archives, Washington D.C.

Price, Laurence. *A Monstrous Shape. Or a Shapelesse Monster. A Description of a female creature borne in Holland, compleat in every part, save only a head like a swine, who hath travailed into many parts, and is now to be seen in London, 1640. Shees loving, courteous, and effeminate, and nere as yet could find a loving mate. "To the tune of the Spanish Pavin"*. London: M.F., [1640?].

Skinker, Tannakin. *A Certaine Relation of the Hog-Faced Gentlewoman called Mistress Tannakin Skinker, who was borne at Wirkham a Neuter Towne betweene the Emperour and the Hollander, scituate on the river Rhyne. Who was betwitched in her mothers wombe in the yeare 1618 and hathe lived ever since unknowne in this kind to any but her Parents and a few other neighbours. And can never recover her true shape, tell she be married, & c. Also relating the cause, as it is since conceived, how her mother came so bewitched*. London: J.O., 1640.

Solomon, Solomon, M.D. *A Guide to Health; or, Advice to Both Sexes. With an Essay on a Certain Disease, Seminal Weakness, and a Destructive Habit of a Private Nature. Also an Address to Parents, Tutors and Guardians of Youth. To Which are Added, Observations on the Use and Abuse of Cold Bathing*. 52nd ed. Stockport, New York: J. Clark, [1800?].

The Universal Spectator and Weekly Journal. "Concerning Gleets and Seminal Weakness of all Kinds." 15 December 1733, n. pag.

—. "Dr Nelson's most wonderful panacea." 15 December 1733, n. pag.

Fiction, Film and Memoir

Anatomy for Beginners. The Anatomists. Dir. Gunther von Hagens, 2005. Madman. DVD.

Coney Island 1940. Promotional film, 1940.

Dunn, Katherine. *Geek Love*. London: Abacus, 2002.

Freaks. Dir. Tod Browning. Warner Brothers, 2004. DVD.

"*Freaks*": *The Sideshow Cinema*. On *Freaks*. Dir. Tod Browning. Warner Brothers, 2004. DVD.

Hall, Sarah. *The Electric Michelangelo*. London: Faber and Faber, 2003.

Juggling Gender. Dir. Tami Gold. Anderson Gold Films, 1992. DVD.

Medical and Other Primary Texts

Acton, William. *Functions and Disorders of the Reproductive Organs*. 4th ed. London: Churchill, 1865.

Adams, Samuel Hopkins. *The Great American Fraud*. New York: n. p., 1906.

Adams, Francis. *The Medical Works of Paulus Aegineta, the Greek Physician, Translated into English with Copious Commentary*. Vol. 1. London: J Welsch, Truettel, Würtz and Co, 1834.

Aretaeus of Cappadocia. *Areteus: Consisting of Eight Books, On the Causes, Symptoms and Cure of Acute and Chronic Diseases*. Trans. John Moffat. London: W. Richardson, 1795.

Barnes, Joseph K., Joseph Janvier Woodward, Charles Smart, George A. Otis and David Lowe Huntingdon. *The Medical and Surgical History of the War of the Rebellion*. Washington, DC: Surgeon-General's Office, 1861–65.

Bartholow, Robert. *Spermatorrhoea: Its Causes, Symptoms, Results and Treatment*. 4th ed. New York: William Wood, 1879.

Beard, George Miller. *American Nervousness: With its Causes and Consequences*. New York: G. P. Putnam's Sons, 1881.

—. *Sexual Neurasthenia: Its Hygiene, Causes and Symptoms and Treatment With a Chapter on Diet for the Nervous*. New York: G. P. Putnam's Sons, 1884.

Bigelow. *Sexual Pathology: A Practical and Popular Review of the Principal Diseases of the Reproductive Organs*. Chicago, IL: Ottaway and Colbert, 1875.

Bourneville, Desiré Magloire and Paul Régnard. *Iconographie photographique de la Salpêtrière*. Paris: Progrès médical, 1877–80.

Brachet, Jean Louis. *Traité de l'hystérie*. Paris: Baillie, 1847.

Briquet, Pierre. *Traité clinique et thérapeutique de l'hystérie*. Paris: J.-B. Baillière, 1859.

The British Medical Journal. Unattributed. 8 Feb, 1873: 151.

Bulwer, John. *Chirologia: or the naturall language of the hand. Composed of the speaking motions, and discoursing gestures thereof. Whereunto is added Chironomia: or, the art of manuall rhetoricke. Consisting of the naturall expressions, digested by art in the hand, as the chiefest instrument of eloquence*. London: J. B. Gent. Philochirosophus, 1644.

Bulwer, John. *Anthropometamorphosis: Man Transform'd; or, the Artificiall Changeling. Historically presented, in the mad and cruel gallantry, foolish bravery, ridiculous beauty, filthy finenesse, and loathsome lovelinesse of most nations, fashioning & altering their bodies from the mould intended by nature; with figures of those transfigurations. To which artificiall and affected deformations are added, all the native and nationall monstrosities that have appeared to disfigure the humane fabrick. With a vindication of the regular beauty and honesty of nature. And an appendix of the pedigree of the English gallant*. London: William Hunt, 1653.

Comstock, Anthony. *Frauds Exposed, or, How the People are Deceived and Robbed, and Youth Corrupted: Being a full exposure of various schemes operated through the mails, and unearthed by the author in a seven years' service as a special agent of the Post Office department and secretary and chief agent of the New York society for the suppression of vice*. New York: J. H. Brown [1880?].

Curling, Thomas. *A Practical Treatise of the Diseases of the Testis, and of the Spermatic Cord and Scrotum.* London: Churchill, 1856.

Dawson, Richard. *An Essay on Spermatorrhea and Urinary Deposits with Observations on the Nature, Causes, and Treatment of the Various Disorders of the Generative System, Illustrated by Cases.* London: Aylott and Jones, 1848.

Dock, Lavinia L. *Hygiene and Morality: A Manual for Nurses and Others, giving an Outline of the Medical, Social, and Legal Aspects of the Venereal Diseases.* New York: G. P. Putnam's Sons, 1910.

Faust, Bernard. *A New Guide to Health: Compiled from the Catechism of Dr. Faust: with Additions and Improvements, Selected from the Writings of Medical Men of Eminence. Designed for the Use of Schools, and Private Families.* Newburyport, MA: W. and J. Gilman, 1810.

First Report of the Committee of the Philadelphia Medical Society on Quack Medicines. Read on the 15th December, 1827, and ordered to be published by the Society. Philadelphia, PA: Judah Dobson, Agent, 1828.

Goldsmith, Alban. *Diseases of the Genito-Urinary Organs.* New York: Wiley and Halsted, 1857.

Gray, Henry. *Gray's Anatomy: Descriptive and Surgical Theory.* New York: Bounty Books, 1977.

Hargreaves, M. K. *Venereal and Generative Diseases including Disorders of Generation, Spermatorrhoea, Prostatorrhoea, Impotence and Sterility in Both Sexes* London: R. Kimpton, 1887.

Harman, Ellen Beard. "Health Convention." *The Herald of Health and Journal of Physical Culture* 5.1 (1865): 8–10.

The Hawkers and Street Dealers of the North of England Manufacturing Districts: including Quack Doctors.–cheap johns.–booksellers by hand… / being some account of their dealings, dodgings, and doings by Felix Folio. Manchester, UK: Abel Heywood, [1858].

Hayes, Albert H. *The Science of Life: Or, Self-Preservation. A Medical Treatise on Nervous and Physical Disability, Spermatorrhoea, Impotence and Sterility.* Boston, MA: Peabody Medical Institute, 1868.

History of the Great Western Sanitary Fair. Cincinnati: C.F. Vent and co., [1884].

Hobbes, Thomas. *Leviathan.* Ed. J. C. A. Gaskin. Oxford and New York: Oxford UP, 1998.

Hollick, Frederick, MD. *The Marriage Guide; Or, Natural History of Generation.* New York: Arno Press, 1974.

—. *A Popular Treatise on Venereal Diseases, in All their Forms. Embracing their History, and Probable Origin; Their Consequences, both to Individuals and to Society; and the Best Modes of Treating Them. Adapted for General Use.* New York: T. W. Strong, 1852.

Hunter, John. *A Treatise on the Venereal Disease.* London: Sherwood, Neely and Jones, 1786.

Hyde, James, and Frank Montgomery. *A Manual of Syphilis and the Venereal Diseases.* Philadelphia, PA: W.B. Saunders, 1895.

Journal of Health. "Quackery." 1.1–4 (1830): 348–49.

—. Unattributed, 1.1 (1830): 262.

Kellogg, John Harvey, M.D. *Plain Facts for Old and Young: Embracing the Natural History of Hygiene of Organic Life.* Burlington, IA: I. F. Segner, 1890.

Lallemand, Claude François. *Des pertes séminales involontaires.* Paris: Béchet jeune, 1836–1842.

—. *A Practical Treatise on the Causes, Symptoms, and Treatment of Spermatorrhoea.* Trans and Ed. Henry J. McDougall. London: John Churchill, 1847.

La'mert, Samuel. *Self-Preservation: A Medical Treatise on the Secret Infirmities and Disorders of the Generative Organs.* London: Samuel La'mert, 1852.

The Lancet. Unattributed, 26 April 1851: 474.

—. Unattributed, 13 April 1853: 156.

—. Unattributed. 15 August 1857: 175.

—. Unattributed. 5 September 1857: 175, 251.

Landouzy, Hector. *Traité complet de l'hystérie*. London and Paris: J. B. Baillière, 1846.

Lydston, Frank. *Sex Hygiene for the Male and What to Say to the Boy*. Chicago, IL: Riverton Press, 1912.

Milton, John. *On Spermatorrhea: Its Pathology, Results, and Complications*. 9th ed. London: Robert Hardwicke, 1872.

—. *Practical Remarks on the Treatment of Spermatorrhoea and Some Forms of Impotence*. London: S. Highley, 1854.

Montaigne, Michel de. *The Complete Essays*. Trans. and ed. M. A. Screech. London and New York: Penguin, 1991.

Paré, A. *Des monstres et prodiges* (1585). Édition critique et commentée par J. Céard. Genève: Librairie Droz, 1971.

Parent-Duchatelet, Alexandre. *De la prostitution dans la ville de Paris: Considérée sous le rapport de l'hygiène publique, de la morale et de l'administration: ouvrage appuyé de documens statistiques puisés dans les archives de la Préfecture de police; précédé d'une notice historique sur la vie et les ouvrages de l'auteur, par Fr. Leuret*. Paris: J.B. Baillière, 1836.

Paris, James du Plessis. *A Short History of Human Prodigies and Monstrous Births, of Dwarfs, Sleepers, Giants, Strong Men, Hermaphrodites, Numerous Births and Extreme Old Age, Etc.* n. p., n. d.

Phillips, Benjamin. "Observations on Seminal and Other Discharges from the Urethra." *London Medical Gazette* (1843): 315–17.

Pickford, Dr. *On True and False Spermatorrhoea*. Trans and Ed. Francis Burdett Courtenay. 3rd ed. London: H. Bailliere, 1854.

Reports of the Medical Society of the City of New-York, on Nostrums, or Secret Medicines: part I. Published by order of the Society, under the direction of the Committee on Quack Remedies. New York: E. Conrad, 1827.

Ricord, Philippe. *Traité pratique des maladies vénériennes: Ou, Recherches critiques et expérimentales sur l'inoculation appliquée à l'étude de ces maladies, suivies d'un résumé thérapeutique et d'un formulaire spécial*. Paris: Rouvier et Le Bouvier, 1838.

Ryan, Michael. *Prostitution in London, with a Comparative View of that of Paris and New York*. London: H. Bailliere, 1839.

Saint-Hilaire, Isidore Geoffroy. *Treatise on Teratology*, 1832.

The Sanitarian: A Monthly Journal 1.1 (April 1873).

Skelton, John. *A Treatise on the Venereal Disease and Spermatorrhoea*. Leeds, UK: Samuel Moxon, 1857.

Tissot, Samuel-Auguste. *Onanism: A Treatise on the Diseases Produced by Masturbation*. Trans. A. Hume. New York: Garland, 1985.

—. *Traité des nerfs et de leurs maladies*. Paris: Didot, 1779.

Vesalius, Andreas. *De Humani Corporis Fabrica*. Basel: J. Oporinus, 1543.

Water-Cure Journal and Herald of Reform: Devoted to Physiology, Hydropathy and the Laws of Life. Vols. 13 and 14. New York: Fowler and Wells, 1852.

Wichmann, Ernest. *De pollutione diurna frequentiori sed rarius observata tabescentiae causa*. Göttingen: J. C. Dieterich, 1782.

Wood-Allen, Mary, MD. *What a Young Girl Ought to Know*. Philadelphia, PA: The Vir Publishing Company, 1905.

Yeoman, T. H. *Debility and Irritability Induced by Spermatorrhoea; the Symptoms, Effects, and Rational Treatment*. London: Effingham Wilson, 1854.

Critical and Secondary Texts

Ackerman, Michael J., Judy Folkenberg and Benjamin Rifkin. *Human Anatomy: Depicting the Body from the Renaissance to Today*. London: Thames and Hudson, 2006.

Adams, Rachel. *Sideshow USA: Freaks and the American Cultural Imagination*. Chicago, IL: U of Chicago P, 2001.

Alberti, Samuel. "Wax Bodies: Art and Anatomy in Victorian Medical Museums." *Museum History Journal*. 2.1 (2009): 7–36.

Altick, Richard. *The Shows of London*. Cambridge, MA: U of Harvard P, 1978.

Armstrong, David, and Elizabeth Metzger Armstrong. *The Great American Medicine Show: Being an Illustrated History of Hucksters, Healers, Health Evangelists, and Heroes from Plymouth Rock to the Prese*. New York: Prentice Hall, 1991.

Austin, Greta. "Marvelous Peoples or Marvelous Races? Race and the Anglo-Saxon *Wonders of the East*." *Marvels, Monsters, and Miracles: Studies in the Medieval and Early Modern Imaginations*. Ed. Timothy S. Jones and David A. Sprunger, 25–51. Kalamazoo, MI: U of Michigan P, 2002.

Bakhtin, Mikhail. *Rabelais and His World*. Trans. Hélène Iswolsky. Bloomington, IN: U of Indiana P, 1984.

Bates, Alan. "Anatomical Venuses: The Aesthetics of Anatomical Modelling in Eighteenth- and Nineteenth-Century Europe." 40th International Congress on the History of Medicine: Proceedings. Ed. J. Pusztai. Budapest: Societas Internationalis Historiae Medicinae, 2006: 183–86.

—. "Dr Kahn's Museum: Obscene Anatomy in Victorian London." *Journal of the Royal Society of Medicine* 99.12 (2006): 618–24.

Beizer, Janet. *Ventriloquized Bodies: Narratives of Hysteria in Nineteenth-Century France*. Ithaca, NY: Cornell UP, 1994.

Bennett, Tony. *The Birth of the Museum: History Theory Politics*. New York: Routledge, 1995.

Benthien, Claudia. *Skin: On the Cultural Border Between Self and the World*. New York: U of Columbia P, 2002.

Berridge, Virginia, and Kelly Loughlin, eds. *Medicine, the Market and the Mass Media: Producing Health in the Twentieth Century*. London: Routledge, 2005.

Bethard, Wayne. *Lotions, Potions, and Deadly Elixirs: Frontier Medicine in America*. Lanham, MD: Taylor Trade, 2004.

Birkett, Jennifer. *The Sins of the Fathers: Decadence in France: 1870–1914*. London: Quartet Books, 1986.

Blon, Philippe, Stephen Bann, Jean-Michel Rey and Jean Louis Schefer, eds. *Voir: la collection Spitzner*. Lagrasse: Editions Verdier, 1988.

Bloom, Michelle. *Waxworks: A Cultural Obsession*. Minneapolis, MN: U of Minnesota P, 2003.

Bogdan, Robert. *Freak Show: Presenting Human Oddities for Amusement and Profit*. Chicago, IL, and London: U of Chicago P, 1988.

Bohde, Daniela. "Skin and the Search for the Interior: The Representation of Flaying in the Art and Anatomy of the Cinquecento." *Bodily Extremes: The Role of Corporeality in the Shaping of Early Modern European Culture and Epistemology*. Ed. Florike Egmond and Robert Zwijnenberg, Aldershot, UK: Ashgate, 2003: 10–47.

Bondeson, Jan. *A Cabinet of Medical Curiosities*. Ithaca and New York: U of Cornell P, 1997.

—. *The Two-Headed Boy and Other Medical Marvels*. Ithaca: U of Cornell P, 2000.

Bosker, Gideon, and Carl Hammer. *Freak Show: Sideshow Banner Art*. San Francisco, CA: Chronicle Books, 1996.

Braidotti, Rosi. *Metamorphoses: Towards a Materialist Theory of Becoming*. Malden, MA: Blackwell Publishers, 2002.

Brock, Pope. *Charlatan: America's Most Dangerous Huckster, the Man Who Pursued Him, and the Age of Flimflam*. New York: Random House, 2008.

Brown, Julie K. *Making Culture Visible: The Public Display of Photography at Fairs, Expositions and Exhibitions in the United States, 1847–1900*. Amsterdam: Harwood Academic Publishers, 2001.

—. *Health and Medicine on Display: International Expositions in the United States, 1876–1904*. Cambridge, MA and London: The MIT Press, 2009.

Brown, Peter. *The Body and Society: Men, Women and Sexual Renunciation in Early Christianity*. London: Faber and Faber, 1990.

Brown, Mackenzie and Charleen Moore. "Gunther von Hagens and *Body Worlds*. Part 1: The Anatomist as Prosektor and Propastiker." *The New Anatomist* 276.1 (2004): 8–14.

— "Gunther von Hagens and Body Worlds. Part 2: The Anatomist as Priest and Prophet." *The New Anatomist* 277.1 (2004): 14–20.

Browne, Anthony. "Body Worlds Visitor Throws Paint in Protest." *The Guardian* March 24 2002: 9.

Burmeister, Maritha Rene. *Popular Anatomical Museums in Nineteenth-Century England*. PhD thesis. Rutgers University, 2000.

Burns, Stanley B. *Skin Pictures: Masterpiece Photographs of Nineteenth Century Dermatology: Selections from the Burns Archive*. New York: Burns Archive Press, 2005.

Burns, William E. "The King's Two Monstrous Bodies: John Bulwer and the English Revolution." *Wonders, Marvels, and Monsters in Early Modern Culture*. Ed. Peter G. Platt, Newark and London: U of Delaware P, 1999: 187–202.

Butler, Judith. *Bodies that Matter: On the Discursive Limits of "Sex"*. New York: Routledge, 1993.

Carlino, Andrea, Deanna Petherbridge and Calude Ritschard. *Corps à vif: art et anatomie*. Genève: Département municipal des affaires culturelles, 1998.

Carr, C. "Circus Minimus: Miller Wows 'Em in the Nabes." *The Village Voice* 43.28 (1998): 59.

Cazort, Mimi, Monique Kornell and K. B. Roberts. *The Ingenious Machine of Nature: Four Centuries of Art and Anatomy Text*. Ottawa: National Gallery of Canada, 1996.

Chemers, Michael M. *Staging Stigma: A Critical Examination of the American Freak Show*. New York: Palgrave Macmillan, 2008.

Clair, Colin. *Human Curiosities*. London, New York and Toronto: Abelard-Schuman, 1968.

Cohen, Jeffrey Jerome, ed. *Monster Theory: Reading Culture*. Minnesota: U of Minnesota P, 1996.

—. *Of Giants: Sex, Monsters and the Middle Ages*. Minneapolis, MN: U of Minnesota P, 1999.

Connolly, Kate. "SS Veteran Sacked by Son as Boss of Corpse Factory." *The Telegraph*, 2 March 2005.

Cook, James W. *The Arts of Deception: Playing With Fraud in the Age of Barnum*. Cambridge, UK and MA: U of Harvard P, 2001.

Crary, Jonathan. *Techniques of the Observer: On Vision and Modernity in the Nineteenth Century*. Cambridge, UK and MA: MIT Press, 1990.

Craton, Lilian. *The Victorian Freak Show: The Significance of Disability and Physical Differences in 19th-Century Fiction*. Amherst, NY: Cambria Press, 2009.

Darby, Robert. *A Surgical Temptation: The Demonization of the Foreskin and the Rise of Circumcision in Britain*. Chicago, IL and London: U of Chicago P, 2005.

—. "Pathologizing Male Sexuality: Lallemand, Spermatorrhea, and the Rise of Circumcision." *Journal of the History of Medicine and Allied Sciences* 60.3 (2005): 283–319.

Darrah, William. *Cartes de Visite in Nineteenth-Century Photography*. Gettysburg, PA: W. C. Darrah, 1981.

Daston, Lorraine J. and Katherine Park. *Wonder and the Orders of Nature, 1150–1750*. New York: Zone Books, 1998.

Davidson, Arnold. *The Emergence of Sexuality: Historical Epistemology and the Formation of Concepts*. Cambridge, MA: U of Harvard P, 2001.

Debord, Guy. *La Société du spectacle*. Paris: Gallimard, 1992.

de Ceglia, Francesco Paolo. 2005. "The Rotten Head, the Disemboweled Woman, the Skinned Man: Body Images From Eighteenth-Century Florentine Wax Modelling." *Journal of Science Communication*. 4 (3): 1–7.

Dennett, Andrea Stulman. *Weird and Wonderful: The Dime Museum in America*. New York: U of New York P, 1997.

Didi-Huberman, Georges. *Invention of Hysteria: Charcot and the Photographic Iconography of the Salpêtrière*. Trans. Alisa Hartz. Cambridge, MA: MIT Press, 2003.

—. *Ouvrir Vénus: Nudité, rêve, cruauté*. Paris: Gallimard, 1999.

Doble, Claire. "What Lies Beneath." *The Sydney Morning Herald*, 2 February 2006.

Dreger, Alice Domurat. *One of Us: Conjoined Twins and the Future of the Normal*. Cambridge, MA, and London: U of Harvard P, 2004.

Edelman, Nicole. *Les métamorphoses de l'hystérique: du début du XIXe siècle à la Grande Guerre*. Paris: Éditions La Découverte, 2003.

Egmond, Florike, and Robert Zwijnenberg, eds. *Bodily Extremes: The Role of Corporeality in the Shaping of Early Modern European Culture and Epistemology*. Aldershot, UK: Ashgate, 2003.

Feher, Michael. "Of Bodies and Technologies." In Hal Foster, ed. *Discussions in Contemporary Culture*. Seattle, WA: Bay Press, 1987: 159–65.

Fiedler, Leslie. *Freaks: Myths and Images of the Secret Self*. New York: Simon and Schuster, 1979.

Foucault, Michel. *Abnormal: Lectures at the Collège de France, 1974–1975*. Ed. Valerio Marchetti and Antonella Salomoni. Trans. Graham Burchell. New York: Picador, 2003.

—. *Discipline and Punish: The Birth of the Prison*. Trans. Alan Sheridan. Harmondsworth: Penguin, 1979.

—. *The History of Sexuality, Volume One. The Will to Knowledge*. Trans. Robert Hurley. Harmondsworth: Penguin, 1981.

Friedman, John Block. *The Monstrous Races in Medieval Art and Thought*. Cambridge and Massachusetts: U of Harvard P, 1981.

Frost, Thomas. *The Old Showmen, and the Old London Fairs*. London: Tinsley Bros., 1874.

Gallagher, Catherine, and Thomas Laqueur. *The Making of the Modern Body: Sexuality and Society in the Nineteenth Century*. Berkeley, CA: U of California P, 1987.

Garland Thomson, Rosemarie. *Extraordinary Bodies: Figuring Physical Disability in American Culture and Literature*. New York: U of Columbia P, 1997.

—, ed. *Freakery: Cultural Spectacles of the Extraordinary Body*. New York: U of New York P, 1996.

Gatens, Moira. *Imaginary Bodies: Ethics, Power and Corporeality*. London: Routledge, 1995.

Gerber, David A. "The 'Careers' of People Exhibited in Freak Shows: The Problem of Volition and Valorization." *Freakery: Cultural Spectacles of the Extraordinary Body*. Ed. Rosemarie Garland Thomson, 38–54. New York: U of New York P, 1996.

Gilman, Sander. *Creating Beauty to Cure the Soul: Race and Psychology in the Shaping of Aesthetic Surgery*. Durham, NC: U of Duke P, 1998.

—. "The Image of the Hysteric." *Hysteria beyond Freud*. Ed. Sander Gilman, Helen King, Roy Porter, G. S. Rousseau and Elaine Showalter. Berkeley, CA: U of California P, 1993.

—. *Picturing Health and Illness: Images of Difference*. Baltimore, MD, and London: U of John Hopkins P, 1995.

Goodall, Jane. "Acting Savage." *Body Show/s: Australian Viewings of Live Performance*. Ed. Peta Tait, 14–28. Amsterdam and Atlanta, GA: Rodopi, 2002.

Grosz, Elizabeth. *Volatile Bodies: Towards a Corporeal Feminism*. Sydney, Australia: Allen and Unwin, 1994.

Hacking, Ian. "The Looping Effect of Human Kinds." in D. Sperber ed. et al. *Causal Cognition: An Interdisciplinary Approach*. Oxford: Oxford UP, 1995. 351–83.

—. "Making Up People." In T. Heller, ed. *Reconstructing Individualism*. Stanford, CA: Stanford UP, 1986: 222–36.

Halberstam, Judith. *Skin Shows: Gothic Horror and the Technology of Monsters*. Durham, NC, and London: Duke UP 1995.

Haley, Bruce. *The Healthy Body in Victorian Culture*. Cambridge, UK and MA: U of Harvard P, 1978.

Hall, Lesley, and Roy Porter. *The Facts of Life: The Creation of Sexual Knowledge in Britain, 1650–1850*. New Haven, CT: U of Yale P, 1995.

Haller, John, and Robin Haller. *The Physician and Sexuality in Victorian America*. Urbana, IL: U of Illinois P, 1974.

Haviland, Thomas N., and Lawrence Charles Parish. "A Brief Account of the Use of Wax Models in the Study of Medicine." *Journal of the History of Medicine and Allied Sciences*, 25 (1970): 52–75.

Hau, Michael. *The Cult of Health and Beauty in Germany: A Social History, 1890–1930*. Chicago, IL: U of Chicago P, 2003.

Haycock, David Boyd, and Patrick Wallis, eds. *Quackery and Commerce in Seventeenth-Century London: The Proprietary Medicine Business of Anthony Daffy*. London: The Wellcome Trust Centre for the History of Medicine at UCL, 2005.

Hillman, David, and Carla Mazzio, eds. *The Body in Parts: Fantasies of Corporeality in Early Modern Europe*. New York and London: Routledge, 1997.

Hodges, Frederick Mansfield. *A History of Spermatorrhoea: The Evolution and Legacy of Medical Conceptualisations of a Venereal Disease and Male Debility in Nineteenth-Century America*. Thesis. University of Oxford, 2000.

Hoffmann, Kathryn. "Sleeping Beauties in the Fairground: The Spitzner, Pedley and Chemisé Exhibits." *Early Popular Visual Culture*, 4.2 (2006): 139–59

Holbrook, Stewart H. *The Golden Age of Quackery*. New York: Macmillan, 1959.

Horn, David. *The Criminal Body: Lombroso and the Anatomy of Deviance*. London: Routledge, 2003.

Hoy, Suellen. *Chasing Dirt: The American Pursuit of Cleanliness*. New York and Oxford: Oxford UP, 1995.

Impey, Oliver, and Arthur MacGregor, eds. *The Origins of Museums: The Cabinet of Curiosities in Sixteenth and Seventeenth-century Europe*. Oxford: Oxford UP, 1985.

Jeffries, Stuart. "The Naked and the Dead." The *Guardian*, 19 March 2002.

Jones, D. G. "Re-inventing Anatomy: The Impact of Plastination on how we see the Human Body." *Clinical Anatomy* 15 (2002): 436–40.

Jones, Greta. *Social Hygiene in Twentieth Century Britain*. London: Croom Helm, 1986.

Jones, Meredith. *Skintight: An Anatomy of Cosmetic Surgery*. Oxford: Berg, 2008.

Jordanova, Ludmilla. *Sexual Visions: Images of Gender in Science and Medicine between the Eighteenth and Twentieth Centuries*. Madison, WI: U of Wisconsin P, 1989.

Kasson, John F. *Amusing the Million: Coney Island at the Turn of the Century*. New York: Hill and Wang, 1978.

Kemp, Martin, and Marina Wallace. *Spectacular Bodies: The Art and Science of the Human Body from Leonardo to Now*. Berkely, CA: U of California P, 2000.

Kirshenblatt-Gimblett, Barbara. *Destination Culture: Tourism, Museums, and Heritage*. Berkeley, CA: U of California P, 1998.

Knoppers, Laura Lunger and Joan B. Landes, eds. *Monstrous Bodies: Political Monstrosities in Early Modern Europe*. Ithaca: Cornell University Press, 2004.

Kroker, Arthur, and Marilouise Kroker. *Body Invaders: Sexuality and the Postmodern Condition*. London: Macmillan Education, 1988.

Kuppers, Petra. *Disability and Contemporary Performance: Bodies on Edge*. New York: Routledge, 2004.

—. "Visions of Anatomy: Exhibitions and Dense Bodies." *Differences: A Journal of Feminist Cultural Studies* 15.3 (2004): 123–56.

Lane, Joan. *A Social History of Medicine: Health, Healing and Disease in England, 1750–1950*. London and New York: Routledge, 2001.

Laqueur, Thomas. *Solitary Sex: A Cultural History of Masturbation*. New York: Zone Books, 2003.

Lecouteux, Claude. *Les Monstres dans la pensée médiévale européenne*. 3rd ed. Paris: Presses de l'Université de Paris-Sorbonne, 1999.

Lee, Denny. "The Nickel Empire Longs To Recapture Its Seedy Glory." *The New York Times*, 16 June 2002.

Leiboff, Marett. "A Beautiful Corpse." *Continuum: Journal of Media and Cultural Studies* 19.2 (2005): 221–37.

Lentz, John. "Revolt of the Freaks." *The Bandwagon* 21.5 (1977): 26–28.

Leroi, Armand Marie. *Mutants: On the Form, Varieties and Errors of the Human Body*. London: Harper Collins, 2003.

Lord, Beth. "Foucault's Museum: Difference, Representation, and Genealogy." *Museum and Society* 4.1 (2006): 1–14.

Lucas, Clay. "Melbourne in Line for Corpse Exhibition." *The Age*, 17 January 2006.

MacDonald, Helen. *Human Remains: Episodes in Human Dissection*. Melbourne, Australia: U of Melbourne P, 2005.

Macfadden, Bernarr. *The Virile Powers of Superb Manhood: How Developed, How Lost, How Regained*. New York: Physical Culture Publishing Company, 1900.

Mason, Michael. *The Making of Victorian Sexuality*. Oxford and New York: Oxford UP, 1994.

Mazer, Sharon. "'She's so fat …' Facing the Fat Lady and Coney Island's Sideshows by the Seashore." *Bodies Out of Bounds: Fatness and Transgression*. Ed. J. Evans Braziel and K. LeBesco. Berkeley, CA: U of California P, 2001.

McCoy, Robert. *Quack: Tales of Medical Fraud from the Museum of Questionable Medical Devices*. Santa Monica, CA: Santa Monica Press, 2000.

McGrath, Roberta. *Seeing Her Sex: Medical Archives and the Female Body*. Manchester, UK: Manchester UP, 2002.

McLaren, Angus. *Impotence: A Cultural History*. Chicago, IL: U of Chicago P, 2007.

McNamara, Brooks. *Step Right Up: An Illustrated History of the American Medicine Show*. Jackson, MS: U of Mississippi P, 1996.

Miller, William. "Abraham Chovet." *The Anatomical Record*. 5.4 (1911): 147–72.

Mitchell, David, and Sharon Snyder. "Exploitations of Embodiment: Born Freak and the Academic Bally Plank." *Disability Studies Quarterly* 25.3 (2005): www.dsq-sds.org.

Mitchell, Michael, ed. *Monsters: Human Freaks in America's Gilded Age: The Photographs of Chas Eisenmann*. Ontario, Canada: ECW Press, 2002.

Mumford, Kevin. "'Lost Manhood' Found: Male Sexual Impotence and Victorian Culture." *Journal of the History of Sexuality* 3 (1992): 33–57.

Nickell, Joe. *Secrets of the Sideshow*. Lexington, KY: U of Kentucky P, 2005.

Oppenheim, Janet. *"Shattered Nerves": Doctors, Patients, and Depression in Victorian England*. New York: Oxford UP, 1991.

Pajot, Stéphane. *De la femme à barbe à l'homme-canon: Phénomènes de cirque et de baraque foraine*. Le Château d'Olonne: Éditions d'Orbestier, 2003.

Park, Katherine. *Secrets of Women: Gender, Generation, and the Origins of Human Dissection*. New York: Zone, 2006.

Pearson, Karl. "Laplace." *Biometrika* 21 (1929): 202–216.

Perry, Ruth. "Colonising the Breast: Sexuality and Maternity in Eighteenth-Century England." *The Journal of the History of Sexuality*, 2.2 (1991): 204–234.

Petherbridge, Deanna, and Ludmilla Jordanova. *The Quick and the Dead: Artists and Anatomy*. Berkeley, CA: U of California P, 1997.

Petropoulos, Thrasy. "Seat at the Autopsy Sideshow." BBC News World Edition. 2002.

Pick, Daniel. *Faces of Degeneration: A European Disorder, 1848–1918*. Cambridge, UK: U of Cambridge P, 1989.

Pivar, David. *Purity and Hygiene: Women, Prostitution, and the "American Plan," 1900–1930*. Westport, CT: Greenwood, 2002.

Platt, Peter G., ed. *Wonders, Marvels, and Monsters in Early Modern Culture*. Newark, DE: U of Delaware P, 1999.

Poggesi, Marta. "The Wax Figure Collection in 'La Specola' in Florence." *Encyclopedia Anatomica: A Complete Collection of Anatomical Waxes*. Ed. Monika Von During. Taschen, 1999.

Poignant, Roslyn. *Professional Savages: Captive Lives and Western Spectacle*. Sydney, Australia: U of New South Wales P, 2004.

Porter, Roy. *Quacks: Fakers and Charlatans in English Medicine*. Stroud and Charleston: Tempus, 2000.

Preciado, Beatrice. *Testo Junkie: Sexe, Drogue et Biopolitique*. Paris: Grasset, 2009.

Puccetti, Azzaroli, ML. "Human Anatomy in Wax During the Florentine Enlightenment." *Italian Journal of Anatomy and Embryology* 102.2 (1997): 77–89.

Py, Christiane, and Cécile Vidart. "Les musées d'anatomie sur les champs de foire." *Actes de la recherche en sciences sociales* 60.1 (1985): 3–10.

Reiss, Benjamin. *The Showman and the Slave: Race, Death, and Memory in Barnum's America*. Cambridge, UK and Massachusetts: U of Harvard P, 2001.

Richardson, Ruth. *Death, Dissection, and the Destitute*. Chicago, IL: U of Chicago P, 2001.

Robertson, Graeme. "Melbourne's Public Anatomical and Anthropological Museums, and the Jordans." *The Medical Journal of Australia*. 1.5 (1956): 164–80.

Rosario, Vernon A. *The Erotic Imagination: French Histories of Perversity*. Oxford and New York: Oxford UP, 1997.

Rosenman, Ellen Bayuk. "Body Doubles: The Spermatorrhea Panic." *Journal of the History of Sexuality* 12.3 (2003): 365–99.

Roth, Michael. "Hysterical Remembering." *Modernism/Modernity* 3.2 (1996): 1–30.

Sappol, Michael. *A Traffic of Dead Bodies: Anatomy and Embodied Social Identity in Nineteenth-Century America*. Princeton, N.J.: U of Princeton P, 2002.

Sawday, Jonathan. *The Body Emblazoned: Dissection and the Human Body in Renaissance Culture*. New York and London: Routledge, 1996.

Scott, David. *Paul Delvaux: Surrealizing the Nude*. London: Reaktion Books, 1992.

Schardt, Peter. "The Freak Might Be You!: An Essay on the History of the Sideshow." In Schneider, Hanspeter. *The Last Sideshow*. London: Dazed, 2004

Schiebinger, Linda. *Nature's Body: Gender in the Making of Modern Science*. Boston: Beacon Press, 1993.

Schnalke, Thomas. *Diseases in Wax: The History of the Medical Moulage*. Trans. Kathy Spatschek. Carol Stream, IL: Quintessence Books, 1995.

Schneider, Hanspeter. *The Last Sideshow*. London: Dazed, 2004.

Schulte-Sasse, Linda. "Advise and Consent: On the Americanization of Body Worlds." *BioSocieties* 1.4 (2006): 369–84.

Schwartz, Vanessa R. *Spectacular Realities: Early Mass Culture in Fin-de-siècle Paris*. Berkeley, CA: U of California P, 1998.

Seigel, Fred. "Theatre of Guts: an Exploration of the Sideshow Aesthetic." *Drama Review*. 35.4 (1991): 107–24.

Seitler, Dana. "Queer Physiognomies; Or, How Many Ways Can We Do the History of Sexuality?" *Criticism* 46.1 (2004): 71–102.

Sekula, Allan. "The Body and the Archive." *October* 39 (1986): 3–64

Sellers, Charles Coleman. *Mr. Peale's Museum: Charles Willson Peale and the First Popular Museum of Natural Science and Art*. New York: Norton, 1980.

Sengoopta, Chandak. "Glandular Politics: Experimental Biology, Clinical Medicine, and Homosexual Emancipation in Fin-de-Siecle Central Europe." *Isis*. 89.3 (1998): 445–73.

Shildrick, Margrit. *Leaky Bodies and Boundaries: Feminism, Postmodernism and (Bio) ethics*. London and New York: Routledge, 1997.

—. *Embodying the Monster: Encounters With the Vulnerable Self*. London; Thousand Oaks, CA: Sage, 2002.

Showalter, Elaine. *Hystories: Hysterical Epidemics and Modern Culture*. New York: U of Columbia P, 1997.

Siegel, Fred. "Theatre of Guts: An Exploration of the Sideshow Aesthetic." *The Drama Review* 35.4 (1991): 107–124.

Simon, Jonathan. "The Theatre of Anatomy: The Anatomical Preparations of Honoré Fragonard." *Eighteenth-Century Studies* 36.1 (2002): 63–79.

Smith, Norman R. "Portentous Births and the Monstrous Imagination in Renaissance Culture." *Marvels, Monsters, and Miracles: Studies in the Medieval and Early Modern Imaginations*. Ed. T. S. Jones and D. A. Sprunger. Kalamazoo, MI: Medieval Institute Publications, 2002: 267–83.

Snow, Robert, and David Wright. "Coney Island: A Case Study In Popular Culture and Technical Change." *Journal of Popular Culture* 9.4 (1976): 960–75.

Solomon-Godeau, Abigail. *Photography at the Dock: Essays on Photographic History, Institutions, and Practices*. Minneapolis, MN: U of Minnesota P, 1991.

Stafford, Barbara. *Body Criticism: Imaging the Unseen in Enlightenment Art and Medicine*. Cambridge, UK and MA: MIT Press, 1991.

Stern, Megan. "Shiny, Happy People: 'Body Worlds' and the Commodification of Health." *Radical Philosophy*, 118 (2003): 2–6.

Sterngass, Jon. *First Resorts: Pursuing Pleasure at Saratoga Springs, Newport and Coney Island*. Baltimore and London: U of Johns Hopkins P, 2001.

Taylor, Gary. *Castration: An Abbreviated History of Western Manhood*. New York and London: Routledge, 2000.

Tort, Patrick. *L'ordre et les monstres: Le débat sur l'origines des déviations anatomiques au XVIIIe siècle*. Paris: Éditions Syllepse, 1998.

Tosh, John. *A Man's Place: Masculinity and the Middle-Class Home in Victorian England*. New Haven, CT, and London: U of Yale P, 1999.

Townsend, J. "The Artistry of Clemente Susini and the La Specola Waxes." *J Biocommun* 27.4 (2000): 2–9.

Tromp, Marlene, ed. *Victorian Freaks: The Social Context of Freakery in Nineteenth-Century Britain*. Columbus, OH: Ohio State UP, 2008.

Van Dijck, José. *The Transparent Body: A Cultural Analysis of Medical Imaging*. Seattle, WA, and London: U of Washington P, 2005.

Verbrugge, Martha H. *Able-Bodied Womanhood: Personal Health and Social Change in Nineteenth-Century Boston*. New York: Oxford UP, 1988.

Vogel, Klaus. "The Transparent Man: Some Comments on the History of a Symbol." *Manifesting Medicine: Bodies and Machines*. Ed. Robert Bud, Bernard Finn and Helmut Trischler. Amsterdam: Harwood Academic Publishers, 1999: 31–62.

Vom Lehn, Dirk. "The Body as Interactive Display: Examining Bodies in a Public Exhibition." *Sociology of Health and Illness* 28.2 (2006): 223–51.

Waldby, Catherine. "The Body and the Digital Archive: the Visible Human Project and the Computerization of Medicine." *Health* 1.2 (1997): 227–43.

Walter, Tony. "Body Worlds: Clinical Detachment and Anatomical Awe." *Sociology of Health and Illness* 26.4 (2004): 464–88.

—. "Plastination for Display: A New Way to Dispose of the Dead." *Journal of the Royal Anthropological Institute* 10 (2004): 603–27.

Warner, Marina. *Fantastic Metamorphoses, Other Worlds*. Oxford: Oxford UP, 2002.

Wegenstein, Bernadette. "Getting Under the Skin, or, How Faces Have Become Obsolete." *Configurations* 10.2 (2002): 221–59.

Werbel, Amy. *Thomas Eakins: Art, Medicine, and Sexuality in Nineteenth-Century Philadelphia*. New Haven, CT: U of Yale P, 2007.

Williams, David. *Deformed Discourse: The Function of the Monster in Mediaeval Thought and Literature*. Exeter, UK: U of Exeter P, 1996.

Witkin, Joel-Peter, ed. *Harms Way: Lust and Madness, Murder and Mayhem*. 2nd ed. Santa Fe, NM: Twin Palms, 1994.

Youngquist, Paul. *Monstrosities: Bodies and British Romanticism*. Minneapolis, MN, and London: U of Minnesota P, 2003.

Index

Amazing Human Body, The (exhibition) 1, 3, 10, 21, 142n7

anatomical museums, commercial, *see* American Museum; Cosmorama Rooms; Dr Jordan and Dr Beck's Anthropological Museum; Egyptian hall; Kahn's Museum of Anatomy; London Anatomical Museum; Musée Quitout; New York Anatomical Gallery; New York Museum of Anatomy; New York Museum of Anatomy, Science and Art; Pacific Museum of Anatomy and Natural Science; Philadelphia Museum of Anatomy; Rackstrow's Museum of Anatomy and Curiosities; Regents Gallery; Spitzner's Grand Musée Anatomique et Ethnologique; Woodhead's Anatomical Museum

anatomical museums, professional, *see* Army Medical Museum (National Museum of Health and Medicine); Hunterian Museum of the Royal College of Surgeons; Museum of the Vienna School of Surgeons (Josephinum); Mütter Museum of the College of Physicians, Philadelphia; National Museum of Health and Medicine (Army Medical Museum); Natural History Museum at the University of Florence (La Specola); Peale's Museum

advertising, medical 2, 14, 15, 19, 20, 24n15, 53, 55–61, 67–9, 71, 128, 141, 147

American Museum 87, 121

American Social Hygiene Association 60, 61, 83, 84n5, 86n16

anatomy, history of 1–22, 23n2, 23n3, 23n4, 24n9, 24n13, 24n14, 125–41, 141n2, 141n3, 142n9

Anatomy Act 4, 16, 37, 40, 50n13, 51n16, 127, 141n3

Aretaeus 65–7, 74, 85n8

Army Medical Museum 6, 18, *see also* National Museum of Health and Medicine

artificial anatomy, history of 24n9, 27, 31–2, 35, 46, 50n14, 51n18, 51n20

Auzoux, Louis 6, 7, 35, 50n12, 52n29

Baartman, Saartjie (the "Hottentot Venus") 23n2, 90, 122n4

Barnum and Bailey Circus 19, 98

Barnum, P. T. 87–8, 120–1, 121n1, 122n2, 122n4, 127, 128, 129, 133

bearded lady 8, 97, 102, 104, 113, 116, 117–19, 121n1

Body Worlds 1–5, 10, 21, 23n1, 23n2, 125–30, 133–7, 140–2, 141n1, 141n2, 142n6, 142n8, 145, *see also* von Hagens, Gunther

Brinkley, John 20, 25n17, 60–2, 84n3, 84n4

Bulwer, John 95–7, 123n16

Burke and Hare 4, 23n5
Byrne, Charles (the "Irish Giant") 8, 24n7, 90

cartes de visite 7, 102, 105, 109
Chang and Eng 90, 102, 147, *see also* conjoined twins
Charcot, Jean-Martin 6, 90, 102, 105–11, 124n25, 124n26
Chovet, Abraham 33–5, 37–9, 43, 48, 50n15, 68, 136
Circus Amok 92, 116
Columbian Exposition (Chicago) 10, *see also* World's Fairs
Coney Island 112, 114, 116
Coney Island Side Show 6, 91–3, 99, 111–20, 122n9
conjoined twins 8, 19, 88, 146, *see also* Chang and Eng; Millie-Christine
Cosmorama Rooms 8, 88
Courney, Martin 6
Crachami, Caroline ("the Sicilian Dwarf") 8, 24n7, 88, 90

degeneration 52n29, 62, 72, 74, 110
Desnoües, Guillaume 32–3, 34, 35, 136
disability rights, activism 98, 99
dissection 4–5, 16, 23n3, 23n4, 24n7, 27, 31, 34–5, 38–9, 42, 46, 50n12, 51n16, 88, 90, 122n7, 127–8, 130, 132–4, 140, 149
Dr Jordan and Dr Beck's Anthropological Museum 57
Dreamland 118, 123n12
dwarves 8, 9, 88, 90, 91, 104, 113, 121n1, *see also* Tom Thumb; Crachami, Caroline

écorché 1, 52n27, 125–6, 129, 130–2, 134–40, 141n1
Egyptian Hall 8, 88
Eisenmann, Charles 90, 102–7, 109–11
ejaculation 64, 66, 74, 85n7
eugenics 6, 86n16

Food and Drug Act 25n18
freaks, professional, *see* Baartman, Saartjie (the "Hottentot Venus"); Byrne, Charles (the "Irish Giant"); Chang and Eng; Crachami, Caroline (the "Sicilian Dwarf"); Fraser, Mat; Heth, Joyce; Jordan, Otis (the "Frog Man"); Koko the Killer Clown; Miller, Jennifer; Millie-Christine; Mills, Fanny; Pastrana, Julia; Tom Thumb; *see also* bearded ladies; conjoined twins; dwarves; giants
Fraser, Mat 113, 114–16, 121

Galen 4, 65, 85n8
Galton, Francis 6, 86n16
Geek Love 91
giants 8–9, 90, 104, 121n1, 124n23, *see also* Byrne, Charles (the "Irish Giant")
gonorrhoea 53, 64–5, 77, 81, 85n8
Great Exhibition, London 6, 10, *see also* World's Fairs
gynaecology 32, 34, 44, 46

health 3, 6, 10–13, 16, 18, 19, 20, 24n9, 24n11, 24n15, 31, 40, 46–9, 53–6, 58, 62–5, 67, 70–1, 73–4, 77, 79, 81, 83, 84n2, 85n9, 104, 124n22, 128–9, 134–5, 137, 141, 144, 146–7, 149
Heth, Joyce 87, 88, 90, 120, 127
"Hottentot Venus", *see* Baartman, Saartjie
Hunter, John 6, 8, 66–7, 71, 76
Hunterian Museum of the Royal College of Surgeons 6, 8, 90, 142n9
hygiene 12, 48, 62, 77, 84n5
Hygiene Exhibition 6
hysteria 73, 90, 102, 106–7, 110–11, 124n25

impotence 62, 68, 72–3, 75, 77, 81–2, 86n13
International Health Exhibition 6,12
"Irish Giant" 8, 90, *see also* Byrne, Charles

Jim Rose Circus Sideshow 92
Jordan, Otis 98–100, 113, 114, 123n19
Jordan, Robert Jacob 14, 56–9, 69, 75, 78, 84n1, 85n9
Josephinum 6, 36, 37, 42, 52n27, *see also* Museum of the Vienna School of Surgeons

Kahn, Joseph 14–16, 24n10, 24n13, 24n14, 51n19, 56, 69, 70–2, 75, 77–8, 82, 85n9

Kahn's Museum of Anatomy 5, 6, 10, 11, 13, 14–17, 24n13, 24n14, 46, 55–6, 72–3, 84n1
Kellogg, John Harvey 20
Koko the Killer Clown 113–16

La Specola 6, 28, 36, 42, 51n18, *see also* Natural History Museum at the University of Florence
Lallemand, Claude François 73–6, 85n11, 86n12, 86n13
London Anatomical Museum 56

Macfadden, Bernarr 63–4, 77, 79, 83
masturbation 52n29, 54, 63–4, 69–70, 73, 77, 108
maternity 31, 43–7, 55, 74, 95, 123n15, 143, *see also* motherhood
Medical Act 16
medical shows 17–18
Miller, Jennifer 113, 116, 117–19, 121
Millie-Christine 96, 147, *see also* conjoined twins
Mills, Fanny 102–5, 107, 109, 124n22, 124n24
motherhood 1, 11, 34, 40, 44–7, *see also* maternity
Musée d'anatomie 8, 26, *see also* Musée Delmas-Rouvière-Ofila
Musée de l'homme 90
Musée Delmas-Rouvière-Ofila 52n25, 85n11, *see also* Musée d'anatomie
Musée Quitout 46, 49n3
Museum of the Vienna School of Surgeons (Josephinum) 6, 36, 37, 42, 52n27
Mütter Museum of the College of Physicians 6, 8, 9, 90, 142n9

National Museum of Health and Medicine (Army Medical Museum) 6, 18, 142n9
Natural History Museum at the University of Florence (La Specola) 6, 28, 36, 42, 51n18
New York Anatomical Gallery 15, 17
New York Museum of Anatomy 57
New York Museum of Anatomy, Science and Art 57

Obscene Publications Act 15–16

Pacific Museum of Anatomy and Natural Science 56
Paré, Ambroise 95–7, 123n14
Pastrana, Julia 88, 89, 122n3
patent medicine 14, 16, 18, 20, 25n18, 54, 56, 57–9, 61, 67, 75
Peale's Museum 14
photography 4, 18–21, 41, 59, 64, 88, 90, 101–11, 123n20, 124n21, 141n1, 148
prostitution 62–3, 84n5, 107
public health 11–12, 62–3, 129
public education campaigns 2, 60
purity movements 62, 83, 84n5

Rackstrow's Museum of Anatomy and Curiosities 5, 35, 51n17
Regent Gallery 8, 88–9
Reinhardt brothers 59, 78
rejuvenation therapy 20, 60, 62, *see also* virility
Ruysch, Frederik 27, 29

semen 64, 66–7, 70, 73–4, 78
seminal loss 55, 64, 66–7, 70–1, 74, 76
seminal weakness 65–9, 71, 76–7
"Sicilian Dwarf", *see* Crachami, Caroline
Solomon, Samuel 19–20, 25n16, 67–8
spermatorrhée 73–4, 76, *see also* spermatorrhoea
spermatorrhoea 53, 55, 69, 71–84, 85n10, 85n12, 86n13, 86n14, 105, 146, *see also* spermatorrhée
Spitzner, Pierre 12, 26, 27, 28, 41, 42, 43, 46, 48, 49n1, 52n25
Spitzner's Grand Musée Anatomique et Ethnologique 5, 6, 8, 10, 11, 13, 26, 42, 47, 49n2, 49n3, 52n24, 52n25
Susini, Clemente 35–7, 41, 42, 44–6, 50n14
syphilis 53, 81

Tissot, Samuel Auguste 64, 84n6
Tom Thumb 88, *see also* dwarves
"transparent man" 6

Valverde 130–2, 136, 140
venereal disease 16, 24n12, 47–8, 52n29, 53, 57, 63, 66–8, 75, 78, 81, 84n5

Vesalius 133–4, 139
virility 20, 60, 62–5, 68, 71, 73–4,
    85n7, 85n10, *see also* rejuvenation
    therapy
von Hagens, Gunther 1–2, 21, 23n1,
    125–40, 141n1, 141n3, 142n5,
    142n7, 142n8, 145, *see also Body
    Worlds*

Wonderland 112, 118
Woodhead's Anatomical Museum 15,
    17
World Exposition (Paris) 10, 19, *see
    also* World's Fairs

World's Fairs
    Chicago 6, 10, *see also* Columbian
    Exposition
    London 6, 10, 17, 41, *see also* Great
    Exhibition
    Paris 10, 19, *see also* World
    Exposition
    New York 87

Zumbo, Gaetano Giulo 27, 28, 30, 31,
    32, 43, 50n14